A Text Book for F.Y.B.Sc./F.Y.B.A.
Semester - II
MATHEMATICS PAPER - II : MT-122 (Credit 2)
Choice Based Credit System (CBCS)(2019 Pattern)

CALCULUS - II

Dr. Shrikisan Gaikwad
M.Sc., B.Ed., M.Phil., Ph.D.
New Arts, Commerce and Science College
Ahmednagar

Dr. Kalyanrao Takale
M.Sc., B.Ed., Ph.D.
RNC Arts, JDB Commerce and NSC Science College
Nashik Road, Nashik

Dr. Pravin Jadhav
M.Sc., Ph.D.
Hon. Balasaheb Jadhav Arts, Commerce & Science College
Ale, Junnar, Pune

Dr. Vikas Jadhav
M.Sc., Ph.D.
Nowrosjee Wadia College
Pune

Prof. Veena P. Kshirsagar
M.Sc., M.Phil., Ph.D.
MIT World Peace University
Kothrud, Pune

Prof. S. R. Patil
M.Sc.
Ex. HOD., S.M. Joshi College
Hadapsar, Pune

Prof. S. M. Waingade
Ex. HOD, Shri. Chhatrapati Shivaji Mahavidyala
Shrigonda, Ahmednagar

N5041

CALCULUS - II   **ISBN 978-93-89686-59-3**

First Edition   :   **December 2019**

©   :   **Authors**

Published By :
NIRALI PRAKASHAN
Abhyudaya Pragati, 1312, Shivaji Nagar
Off J.M. Road, PUNE – 411005
Tel - (020) 25512336/37/39, Fax - (020) 25511379
Email : niralipune@pragationline.com

➢ **DISTRIBUTION CENTRES**

PUNE

Nirali Prakashan : 119, Budhwar Peth, Jogeshwari Mandir Lane, Pune 411002, Maharashtra

(For orders within Pune)   Tel : (020) 2445 2044, Mobile : 9657703145

Email : niralilocal@pragationline.com

Nirali Prakashan : S. No. 28/27, Dhayari, Near Asian College Pune 411041

(For orders outside Pune)   Tel : (020) 24690204; Mobile : 9657703143

Email : bookorder@pragationline.com

MUMBAI

Nirali Prakashan : 385, S.V.P. Road, Rasdhara Co-op. Hsg. Society Ltd.,

Girgaum, Mumbai 400004, Maharashtra; Mobile : 9320129587

Tel : (022) 2385 6339 / 2386 9976, Fax : (022) 2386 9976

Email : niralimumbai@pragationline.com

➢ **DISTRIBUTION BRANCHES**

JALGAON

Nirali Prakashan : 34, V. V. Golani Market, Navi Peth, Jalgaon 425001, Maharashtra,

Tel : (0257) 222 0395, Mob : 94234 91860; Email : niralijalgaon@pragationline.com

KOLHAPUR

Nirali Prakashan : New Mahadvar Road, Kedar Plaza, 1st Floor Opp. IDBI Bank, Kolhapur 416 012

Maharashtra. Mob : 9850046155; Email : niralikolhapur@pragationline.com

NAGPUR

Nirali Prakashan : Above Maratha Mandir, Shop No. 3, First Floor,

Rani Jhanshi Square, Sitabuldi, Nagpur 440012, Maharashtra

Tel : (0712) 254 7129; Email : niralinagpur@pragationline.com

DELHI

Nirali Prakashan : 4593/15, Basement, Agarwal Lane, Ansari Road, Daryaganj

Near Times of India Building, New Delhi 110002 Mob : 08505972553

Email : niralidelhi@pragationline.com

BENGALURU

Nirali Prakashan : Maitri Ground Floor, Jaya Apartments, No. 99, 6th Cross, 6th Main,

Malleswaram, Bengaluru 560003, Karnataka; Mob : 9449043034

Email: niralibangalore@pragationline.com

Other Branches : Gujarat, Kolkata Hyderabad, Chennai

niralipune@pragationline.com | www.pragationline.com

Also find us on   www.facebook.com/niralibooks

Preface

This book is based on a course Calculus-II. The purpose of this text book is to provide a rigorous treatment of the foundations of differential calculus. We write this book as per the revised syllabus of F.Y. B.Sc. Mathematics, revised by Savitribai Phule Pune University, Pune, implemented from June 2019. Calculus is the most useful subject in all of mathematics and it is used extensively in applied mathematics and engineering.

In Chapter 1, we develop the theory of differentiation of real valued function of single variable. Furthermore, we study algebraic properties of derivative. We prove Interior extremum theorem, Mean Value theorems and their Consequences, Intervals of increasing and decreasing of a function, first derivative test for extrema and as an applications of Mean Value theorems we solved some examples. The Chapter 2, is devoted for Indeterminate forms, L' Hospital Rules, Taylor theorem and Maclaurin theorem with Lagranges form of remainder. Also, we study Successive Differentiation and obtain the nth derivatives of functions. We proved Leibnitz theorem for successive differentiation and solve some examples.

Third Chapter deals with the concept of Ordinary Differential Equations. In this chapter, we study Linear first order equations, Separable equations, Existence and Uniqueness of solutions of nonlinear equations and solve some examples.

In fourth Chapter, we study the Exact Differential Equations. Also, we discuss Transformation of nonlinear equations to separable equations. Furthermore, we study Exact differential equations, Integrating factors and solve some examples.

We are very much thankful to **Mr. Dinesh Furia** and **Mr. Jignesh Furia**, Nirali Prakashan, Pune, for valuable cooperation and guidance. Also, we are specially thankful to Mrs. Nanda Takale for her valuable cooperation. All authors devote this book to their parents.

We welcome any opinions and suggestions which will improve the future editions and help readers in future.

In case of queries/suggestions, send an email to: *kalyanraotakale@rediffmail.com*

- Authors

Syllabus: PAPER-II: MT-122: CALCULUS - II

1. Unit 1: Differentiation [10]

   1.1 The Derivatives:

   Definition of the derivative of a function at a point, every differentiable function is continuous, Rules of differentiation, Caratheodary's theorem(without proof), The chain rule, Derivative of inverse function (without proof , only examples).

   1.2 The Mean Value Theorems:

   Interior extremum theorem, Mean Value theorems and their Consequences, Intervals of increasing and decreasing of a function, first derivative test for extrema.

2. Unit 2: L Hospital Rule and Successive Differentiation [10]

   2.1 L Hospital Rule:

   Indeterminate forms, L Hospital Rules (without proof).

   2.2 Taylor's theorem:

   Taylors theorem and Maclaurins theorem with Lagranges form of remainder (Without proof).

   2.3 Successive Differentiation:

   The nth derivative and Leibnitz theorem for successive differentiation.

3. Unit 3: Ordinary Differential Equations [08]

   3.1 Linear first order equations.

   3.2 Separable equations.

   3.3 Existence and Uniqueness of solutions of nonlinear equations.

4. Unit 4: Exact Differential Equations [08]

   4.1 Transformation of nonlinear equations to separable equations.

   4.2 Exact differential equations.

   4.3 Integrating factors.

Contents

Chapter 1

Unit 1: Differentiation

1.1   Introduction

In the seventeenth century, European mathematicians Isaac Barrow, Ren Descartes, Pierre de Fermat, Blaise Pascal, John Wallis and others discussed the idea of a derivative. Calculus was developed in the latter half of the seventeenth century by two mathematicians, Gottfried Leibniz and Isaac Newton. Newton first developed calculus and applied it directly to the understanding of physical systems. At the same time, independently, Leibniz developed the notations used in calculus. Differential calculus determines the rate of change of a quantity. The primary objects of study in differential calculus

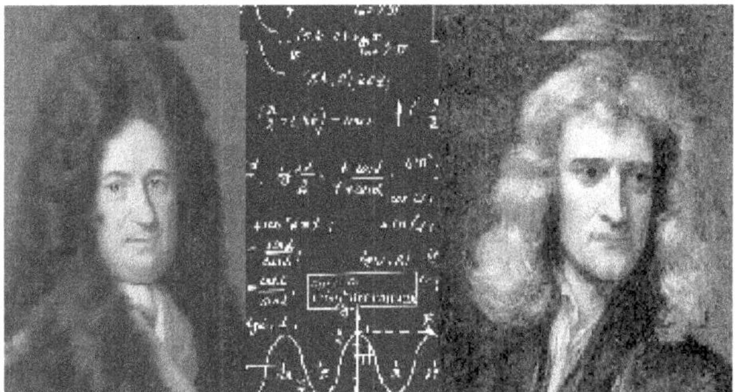

Figure 1.1: Gottfried Leibniz and Isaac Newton

are the derivative of a function, related notions such as the differential, and their applications. The derivative of a function at a chosen input value describes the rate of change of the function near that input value. The process of finding a derivative is called differentiation. Calculus has many practical applications in real life. Some of the concepts that use calculus include motion, electricity, heat, light, harmonics, acoustics, and astronomy. Calculus is used in geography, computer vision (such as for autonomous driving of cars), photography, artificial intelligence, robotics, video games, and even movies. Also,economists use calculus to predict supply, demand, and maximum potential profits.

1.2   The Derivative

In this section we will present some of the elementary properties of the derivative. We begin with the definition of the derivative of a function.

Definition 1.1. *Suppose f is a (real valued) function defined on an open interval I and $c \in I$. Then f is said to be differentiable at c if*

$$\lim_{x \to c} \frac{f(x) - f(c)}{x - c}$$

exists, and in that case the value of the limit is called the derivative of f at c.

The derivative of f at c, if exists, is denoted by

$$f'(c) \ or \ \left(\frac{df}{dx}\right)_{x=c}.$$

If we put $x - c = h$ then a function f is called differentiable at $x = c$ if

$$\lim_{h \to 0} \frac{f(c+h) - f(c)}{h}$$

exists. Therefore,

$$f'(c) = \lim_{h \to 0} \frac{f(c+h) - f(c)}{h}.$$

Graphically, this definition says that the derivative of f at c is the slope of the tangent line to $y = f(x)$ at c, which is the limit as $h \to 0$ of the slopes of the lines through $(c, f(c))$ and $(c + h, f(c + h))$.

Remark 1.1. *Next, we consider some examples of non-differentiability at discontinuities, corners, and cusps. A function can fail to be differentiable at point if:*

1. The function is not continuous at the point. The function $f : \mathbb{R} \to \mathbb{R}$ defined by

$$f(x) = \begin{cases} \frac{|x|}{x}, & if \ x \neq 0; \\ 0, & if \ x = 0. \end{cases}$$

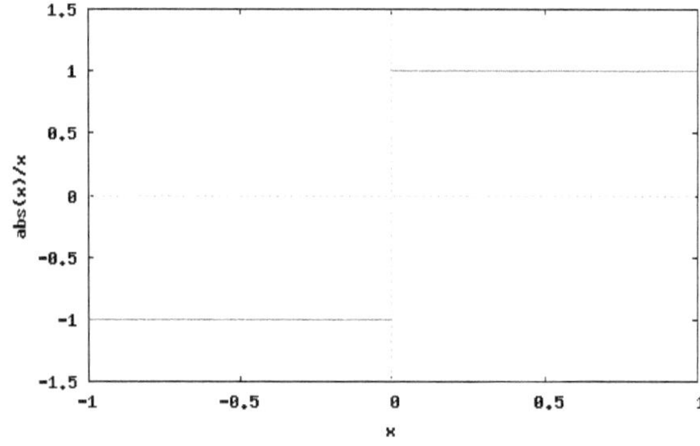

Figure 1.2

2. If f has a sharp corner at x. The absolute value function $f(x) = |x|$ is not differentiable at $x = 0$.

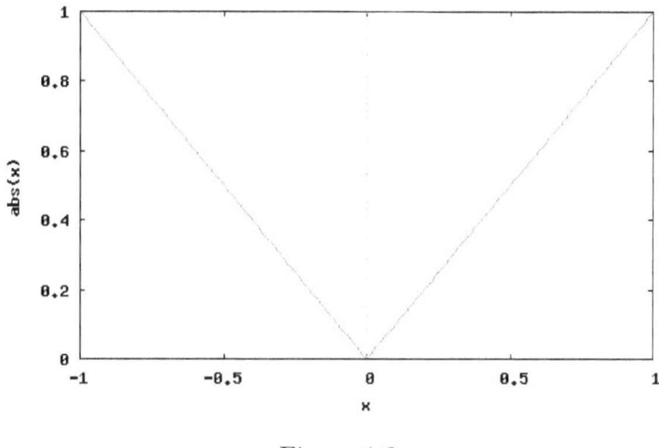

Figure 1.3

3. If f has a vertical tangent line at x (since the derivative is the slope of the tangent line and the slope of a vertical line is undefined). The function $f : \mathbb{R} \to \mathbb{R}$ defined by $f(x) = x^{1/3}$.

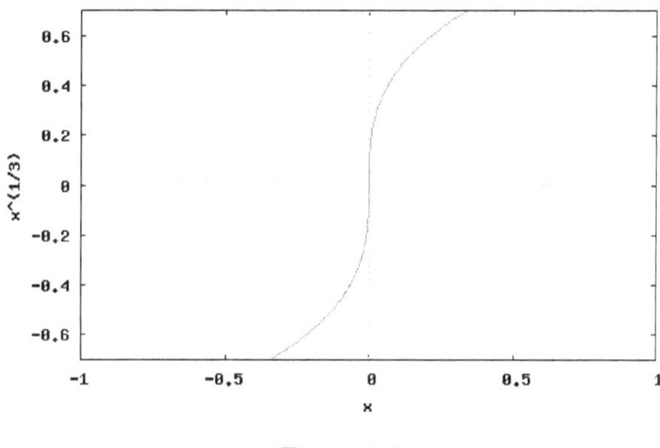

Figure 1.4

Example 1.1. *The function $f : \mathbb{R} \to \mathbb{R}$ defined by $f(x) = x^2$ is differentiable on $\mathbb{R}$ with derivative* $f'(x) = 2x$.

Solution: Since, the derivative of f is

$$f'(c) = \lim_{h \to 0} \frac{f(x+h) - f(x)}{h} = \lim_{h \to 0} \frac{(x+h)^2 - x^2}{h}$$

$$= \lim_{h \to 0} \frac{h(2x+h)}{h} = \lim_{h \to 0}(2x + h) = 2x.$$

Therefore, $f(x)$ is differentiable on $\mathbb{R}$ with derivative $f'(x) = 2x$.

Example 1.2. *The function $f : \mathbb{R} \to \mathbb{R}$ defined by*

$$(\ ) = \begin{cases} x^2, & if \ x > 0; \\ \end{cases}$$

is differentiable on $\mathbb{R}$ with derivative

$$f'(x) = \begin{cases} 2x, & if\ x > 0; \\ 0, & if\ x \leq 0. \end{cases}$$

Solution: For $x > 0$, the derivative is $f'(x) = 2x$ as above, and for $x < 0$, we have $f'(x) = 0$.

For $x = 0$, we evaluate the left hand limit and right hand limit as follows

$$\lim_{h \to 0^+} \frac{f(0+h) - f(0)}{h} = \lim_{h \to 0^+} \frac{f(h)}{h}$$
$$= \lim_{h \to 0^+} \frac{(0+h)^2 - 0^2}{h}$$
$$= \lim_{h \to 0^+} h$$
$$= 0.$$

$$\lim_{h \to 0^-} \frac{f(0+h) - f(0)}{h} = \lim_{h \to 0^-} \frac{f(h)}{h}$$
$$= 0.$$

Since the left and right limits exist and are equal, the limit also exists, and f is differentiable at $x = 0$ with $f'(0) = 0$.

Example 1.3. *The function $f : \mathbb{R} \to \mathbb{R}$ defined by*

$$f(x) = \begin{cases} \dfrac{1}{x}, & if\ x \neq 0; \\ 0, & if\ x = 0. \end{cases}$$

is differentiable at $x \neq 0$ with derivative $f'(x) = \dfrac{-1}{x^2}$.

Solution: Since

$$\lim_{h \to 0} \frac{f(x+h) - f(x)}{h} = \lim_{h \to 0} \frac{\dfrac{1}{x+h} - \dfrac{1}{x}}{h}$$
$$= \lim_{h \to} \frac{x - (x+h)}{hx(x+h)}$$
$$= \lim_{h \to 0} \frac{-1}{x(x+h)}$$
$$= \frac{-1}{x^2}.$$

However, f is not differentiable at $x = 0$ since the limit

$$\lim_{h \to 0} \frac{f(0+h) - f(0)}{h} = \lim_{h \to 0} \frac{f(h) - 0}{h}$$
$$= \lim_{h \to 0} \frac{\dfrac{1}{h} - 0}{h}$$
$$= \lim_{h \to 0} \frac{1}{h^2}$$

Example 1.4. *The absolute value function $f(x) = |x|$ is differentiable at $x \neq 0$ but it is not differentiable at $x = 0$.*

Solution: Here $f(x) = |x|$.

$$\lim_{h \to 0^+} \frac{f(0+h) - f(0)}{h} = \lim_{h \to 0^+} \frac{f(h)}{h}$$
$$= \lim_{h \to 0^+} \frac{|h| - 0}{h}$$
$$= \lim_{h \to 0^+} \frac{h}{h}$$
$$= 1.$$

$$\lim_{h \to 0^-} \frac{f(0+h) - f(0)}{h} = \lim_{h \to 0^-} \frac{f(h)}{h}$$
$$= \lim_{h \to 0^-} \frac{|h| - 0}{h}$$
$$= \lim_{h \to 0^-} \frac{-h}{h}$$
$$= -1.$$

Since the left and right limits exist but not equal. Thus, $f(x)$ is differentiable at $x \neq 0$ but it is not differentiable at $x = 0$.

Example 1.5. *The function $f : \mathbb{R} \to \mathbb{R}$ defined by*

$$f(x) = \begin{cases} x^2 \sin\left(\dfrac{1}{x}\right), & if\ x \neq 0; \\ 0, & if\ x = 0. \end{cases}$$

Show that f is differentiable on $\mathbb{R}$.

Solution: The function f is differentiable on $\mathbb{R}$. It follows from the product rule given below that f is differentiable at $x \neq 0$ with derivative

$$f'(x) = 2x \sin\left(\frac{1}{x}\right) - \cos\left(\frac{1}{x}\right).$$

Moreover, f is differentiable at $x = 0$ with $f'(0) = 0$, since

$$\lim_{h \to 0} \frac{f(0+h) - f(0)}{h} = \lim_{h \to 0} h\ \sin\left(\frac{1}{h}\right) = 0.$$

1.3 Left and right derivatives

. In many cases, we are using derivatives that are defined only at the interior points of the domain of a function. Now, sometimes, however, it is convenient to use one-sided left or right derivatives that are defined at the endpoint of an interval.

Definition 1.2. *Suppose that $f : [a, b] \to \mathbb{R}$. Then f is said to be right-differentiable at $a \leq c < b$ with right derivative $f'_+(c)$ if*

$$f(c + h) - f(c)$$

Also, f is said to be left-differentiable at $a < c \leq b$ with right derivative $f'_-(c)$ if

$$f'_-(c) = \lim_{h \to 0^-} \frac{f(c+h) - f(c)}{h} \quad \textit{exists.}$$

A function f is differentiable at $a < c < b$ if and only if the left and right derivatives at c both exist and are equal that is $f'_+(c) = f'_-(c)$.

Example 1.6. *The absolute value function $f(x) = |x|$ is not differentiable at $x = 0$.*

Solution: Here $f(x) = |x|$

$$\begin{aligned}
f'_+(0) &= \lim_{h \to 0^+} \frac{f(0+h) - f(0)}{h} \\
&= \lim_{h \to 0^+} \frac{f(h)}{h} = \lim_{h \to 0^+} \frac{|h| - 0}{h} \\
&= \lim_{h \to 0^+} \frac{h}{h} \\
&= 1. \\
f'_-(0) &= \lim_{h \to 0^-} \frac{f(0+h) - f(0)}{h} \\
&= \lim_{h \to 0^-} \frac{f(h)}{h} = \lim_{h \to 0^-} \frac{|h| - 0}{h} \\
&= \lim_{h \to 0^-} \frac{-h}{h} \\
&= -1.
\end{aligned}$$

Thus, $f'_+(c) \neq f'_-(c)$, hence f is not differentiable at $x = 0$.

1.4   Properties of the derivative

In this section, we prove some basic properties of differentiable functions.

Theorem 1.1. *If $f : (a, b) \to \mathbb{R}$ is differentiable at at $c \in (a, b)$, then f is continuous at c.*

Proof: For all $x \in I; x \neq c$, we have

$$f(x) - f(c) = \left(\frac{f(x) - f(c)}{x - c} \right)(x - c)$$

Since f is differentiable at $x = c$, therefore, $f'(c)$ exists.

$$\begin{aligned}
\lim_{x \to c}(f(x) - f(c)) &= \lim_{x \to c} \left(\frac{f(x) - f(c)}{x - c} \right) \left(\lim_{x \to c}(x - c) \right) \\
&= f'(c).0 \\
&= 0.
\end{aligned}$$

Therefore, $\lim_{x \to c} f(x) = f(c)$, so that f is continuous at c.

For example, the absolute value function $f(x) = |x|$ is continuous but not differentiable at $x = 0$.

Note: We observe that continuity of f at a point c is a necessary (but not sufficient) condition for the existence of the derivative at c.

Theorem 1.2. *Let $I \subseteq \mathbb{R}$ be an interval, let $c \in I$, and let $f : I \to \mathbb{R}$ and $g : I \to \mathbb{R}$ be functions that are differentiable at c. Then we prove the following properties.*

(i) If $\alpha \in \mathbb{R}$, then the function αf is differentiable at c, and $(\alpha f)'(c) = \alpha f'(c)$.

(ii) The function $f + g$ is differentiable at c, and $(f + g)'(c) = f'(c) + g'(c)$.

(iii) The function fg is differentiable at c, and $(fg)'(c) = f'(c)g(c) + f(c)g'(c)$.

(iv) If $g(c) \neq 0$, then the function f/g is differentiable at c. and $\left(\frac{f}{g}\right)'(c) = \dfrac{f'(c)g(c) - f(c)g'(c)}{(g(c))^2}$.

Proof: Let $c \in I$, and $f : I \to \mathbb{R}$ and $g : I \to \mathbb{R}$ be functions and that are differentiable at c.

Therefore, $\lim\limits_{x \to c} \dfrac{f(x) - f(c)}{x - c}$ exists.

$$\therefore f'(c) = \lim_{x \to c} \frac{f(x) - f(c)}{x - c}.$$

Also, $\lim\limits_{x \to c} \dfrac{g(x) - g(c)}{x - c}$ exists.

$$\therefore g'(c) = \lim_{x \to c} \frac{g(x) - g(c)}{x - c}.$$

(iii) For $x \in I; x \neq c$, we have

$$\lim_{x \to c} \frac{(fg)(x) - (fg)(c)}{x - c} = \lim_{x \to c} \frac{f(x)g(x) - f(c)g(c)}{x - c}$$

$$= \lim_{x \to c} \frac{f(x)g(x) - f(c)g(x) + f(c)g(x) - f(c)g(c)}{x - c}$$

$$= \lim_{x \to c} \frac{f(x) - f(c)}{x - c} \cdot g(x) + f(c) \lim_{x \to c} \frac{g(x) - g(c)}{x - c}.$$

Now, g is differentiable at c therefore, g is continuous at c, then $\lim\limits_{x \to c} g(x) = g(c)$.

Since f and g are differentiable at c therefore, we conclude that

$$\lim_{x \to c} \frac{(fg)(x) - (fg)(c)}{x - c} = f'(c)g(c) + f(c)g'(c).$$

$$\therefore (fg)' = f'(c)g(c) + f(c)g'(c).$$

Hence, fg is differentiable at c and $(fg)' = f'(c)g(c) + f(c)g'(c)$ is proved.

(iv) Now, g is differentiable at c, therefore, it is continuous at that point. Also, $g(c) \neq 0$, therefore, there exists an interval $J \subseteq I$ with $c \in J$ such that $g(x) \neq 0$ for all $x \in J$.

For $x \in J$, $x \neq c$, we have

$$\lim_{x \to c} \frac{\left(\frac{f}{g}\right)(x) - \left(\frac{f}{g}\right)(c)}{x - c} = \lim_{x \to c} \frac{\frac{f(x)}{g(x)} - \frac{f(c)}{g(c)}}{x - c}$$

$$= \lim_{x \to c} \frac{f(x)g(c) - f(c)g(x)}{g(x)g(c)(x - c)}$$

$$= \lim_{x \to c} \frac{f(x)g(c) - f(c)g(c) + f(c)g(c) - f(c)g(x)}{g(x)g(c)(x - c)}$$

$$= \lim_{x \to c} \frac{1}{g(x)g(c)} \left[\lim_{x \to c} \frac{f(x) - f(c)}{x - c} \cdot g(c) - f(c) \cdot \lim_{x \to c} \frac{g(x) - g(c)}{x - c} \right].$$

Using the continuity of g at c and the differentiability of f and g at c, we get

$$\left(\frac{f}{g}\right)'(c) = \lim_{x \to c} \frac{\left(\frac{f}{g}\right)(x) - \left(\frac{f}{g}\right)(c)}{x - c}$$

$$= \frac{f'(c)g(c) - f(c)g'(c)}{(g(c))^2}.$$

Hence, f/g is differentiable at c and $\left(\dfrac{f}{g}\right)'(c) = \dfrac{f'(c)g(c) - f(c)g'(c)}{(g(c))^2}$ is proved.

Corollary 1.1. *If $f_1, f_2, \cdots, f_n$ are functions on an interval I to $\mathbb{R}$ that are differentiable at $c \in I$, then we have the following properties.*

(i) The function $f_1 + f_2 + \cdots + f_n$ is differentiable at c and

$$(f_1 + f_2 + \cdots + f_n)'(c) = f_1'(c) + f_2'(c) + \cdots + f_n'(c).$$

(ii) The function $f_1 f_2 \cdots f_n$ is differentiable at c, and

$$(f_1 f_2 \cdots f_n)'(c) = f_1'(c)f_2(c) \cdots f_n(c) + f_1(c)f_2'(c) \cdots f_n(c) + \cdots + f_1(c)f_2(c) \cdots f_n'(c).$$

An important special case of the extended product rule occurs if the functions are equal, that is, $f_1 = f_2 = \cdots = f_n = f$. Then $(f^n)'(c) = n(f(c))^{n-1}f'(c)$.

Theorem 1.3. *Carathéodory's Theorem*

Let f be defined on an interval I containing the point c. Then f is differentiable at c if and only if there exists a function φ on I that is continuous at c and satisfies $f(x) - f(c) = \varphi(x)(x - c)$ for $x \in I$. In this case, we have $\varphi(c) = f'(c)$.

To illustrate Carathéodorys theorem, we consider the function f defined by $f(x) := x^3$ for $x \in \mathbb{R}$. For $c \in \mathbb{R}$, we have from factorization

$$x^3 - c^3 = (x^2 + cx + c^2)(x - c)$$

that $\varphi(x) := x^2 + cx + c^2$ satisfies the conditions of the theorem $f(x) - f(c) = \varphi(x)(x - c)$.

Theorem 1.4. Chain Rule

Let I, J be intervals in $\mathbb{R}$, let $g : I \to \mathbb{R}$ and $f : J \to \mathbb{R}$ be functions such that $f(J) \subseteq I$, and let $c \in J$. If f is differentiable at c and if g is differentiable at $f(c)$, then the composite function $g o f$ is differentiable at c and $(g o f)'(c) = g'(f(c)) \cdot f'(c)$.

Proof. Now, the function f is differentiable at $x = c$, therefore, $f'(c)$ exists.

Therefore, by Carathéodorys theorem (3.7), there exists a function φ on J such that φ is continuous at c and

$$f(x) - f(c) = \varphi(x)(x - c) \ for \ x \in \ J, \tag{1.1}$$

where $\varphi(c) = f'(c)$.

Also, g is differentiable at $f(c)$, therefore, $g'(f(c))$ exists.

Therefore, by Carathéodorys theorem (3.7), there exists a function Ψ defined on I such that Ψ is continuous at $d = f(c)$ and

$$g(y) - g(d) = \Psi(y)(y - d) \ for \ y \in \ I,$$

where $\Psi(d) = g'(d)$.

After substituting $y = f(x)$, $d = f(c)$ and by equation(3.1), we get

$$g(f(x)) - g(f(c)) = \Psi(f(x))(f(x) - f(c))$$
$$= [(\Psi o f)(x) \cdot \varphi(x)](x - c)$$
$$\therefore (g o f)(x) = [(\Psi o f)(x) \cdot \varphi(x)](x - c)$$

for all $x \in J$ such that $f(x) \in I$.

Since, the function $(\Psi o f) \cdot \varphi$ is continuous at c and its value at c is $g'(f(c)) \cdot f'(c)$, Carathéodorys theorem gives

$$(g o f)'(c) = g'(f(c)) \cdot f'(c).$$

Note: If g is differentiable on I, if f is differentiable on J, and if $f(J) \subseteq I$, then it follows from the Chain Rule that $(g o f)' = (g' o f) \cdot f'$, which can also be written in the form $D(g o f) = (D g o f) \cdot D f$.

Remark 1.2. *From Chain Rule, we have*

$$(g o f)'(x) = g'(f(x)) \cdot f'(x) \ \ for \ \ x \in I.$$

Therefore, we have rule $(f^n)'(x) = n(f(x))^{n-1} f'(x)$ for all $x \in I$.

1.5 Inverse Functions

In this section, we discuss the derivative of the inverse function, when this inverse function exists.

theorem to ensure the existence of a continuous inverse function.

If f is a continuous strictly monotone function on an interval I, then its inverse function $g = f^{-1}$ is defined on the interval $J := f(I)$ and satisfies the relation

$$g(f(x)) = x \text{ for } x \in I.$$

If $c \in I$ and $d := f(c)$, and if we knew that both $f'(c)$ and $g'(d)$ exist, then we could differentiate both sides of the equation and apply the Chain Rule to the left side to get

$$g'(f(c)) \cdot f'(c) = 1.$$

Thus, if $f'(c) \neq 0$ we would obtain

$$g'(d) = \frac{1}{f'(c)}.$$

However, it is necessary to deduce the differentiability of the inverse function g from the assumed differentiability of f earlier we discussed. This is nicely accomplished by using Carathéodorys theorem.

Theorem 1.5. *Let I be an interval in $\mathbb{R}$ and let $f : I \to \mathbb{R}$ be strictly monotone and continuous on I. Let $J := f(I)$ and let $g : J \to \mathbb{R}$ be the strictly monotone and continuous function inverse to f. If f is differentiable at $c \in I$ and $f'(c) \neq 0$, then g is differentiable at $d := f(c)$ and*

$$g'(d) = \frac{1}{f'(c)} = \frac{1}{f'(g(d))}.$$

Proof. Given $c \in \mathbb{R}$, we obtain from Carathéodorys theorem (3.7) a function φ on I with properties that φ is continuous at c;

$$f(x) - f(c) = \varphi(x)(x - c) \ for \ x \in I \text{ and} \varphi(c) = f'(c).$$

Since $\varphi(c) \neq 0$ by hypothesis, there exists a neighborhood $V := (c - \delta, c + \delta)$ such that $\varphi(x) \neq 0$ for all $x \in V \cap I$.

If $U := f(V \cap I)$, then the inverse function g satisfies $f(g(y)) = y$ for all $y \in U$, so that

$$y - d = f(g(y)) - f(c) = \varphi(g(y)) \cdot (g(y) - g(d)).$$

Since $\varphi(g(y)) \neq 0$ for $y \in U$, we can divide to get

$$g(y) - g(d) = \frac{1}{\varphi(g(y))} \cdot (y - d).$$

Since the function $1/(\varphi o g)$ is continuous at d, we apply Theorem 6.1.5 to conclude that $g'(d)$ exists and

$$g'(d) = 1/\varphi(g(d)) = 1/\varphi(c) = 1/f'(c).$$

Theorem 1.6. *Let I be an interval and let $f : I \to \mathbb{R}$ be strictly monotone on I. Let $J := f(I)$ and let $g : J \to \mathbb{R}$ be the function inverse to f. If f is differentiable on I and $f'(x) \neq 0$ for $x \in I$, then g is differentiable on J and*

$$g' = \frac{1}{f' o g}$$

Example 1.7. *The function $f : \mathbb{R} \to \mathbb{R}$ defined by*

$$f(x) := x^5 + 4x + 3$$

is continuous and strictly monotone increasing (since it is the sum of two strictly increasing functions). Moreover, $f'(x) = 5x^4 + 4$ is never zero. Therefore, by theorem (3.31), the inverse function $g = f^{-1}$ is differentiable at every point.

If we take $c = 1$, then $f(1) = 8$.

Therefore, we obtain $g'(8) = g'(f(1)) = 1/f'(1) = 1/9$.

Example 1.8. *The inverse of the function $f(x) = x^2$ with reduced domain $[0, \infty)$ is $f^{-1}(x) = \sqrt{x}$. Use the formula given above to find the derivative of $f^{-1}(x)$.*

Solution: We have $f(x) = x^2$

$$f'(x) = 2x, \quad f'(f^{-1}(x)) = 2\sqrt{x}.$$

Using the formula above, we have

$$(f^{-1})'(x) = \frac{1}{f'(f^{-1}(x))} = \frac{1}{2\sqrt{x}}.$$

Illustrative Examples

1. Use the definition to find the derivative of each of the following functions:

   (a) $f(x) = x^3$ for $x \in \mathbb{R}$,   (b) $g(x) = 1/x$ for $x \in \mathbb{R}, x \neq 0$

   (c) $h(x) = \sqrt{x}$ for $x > 0$,   (d) $k(x) = 1/\sqrt{x}$ for $x > 0$.

   Solution: (a) Let $f := x^3$ for $x \in \mathbb{R}$. For $c \in \mathbb{R}$, by definition of derivative, we have

$$f'(c) = \lim_{x \to c} \frac{f(x) - f(c)}{x - c}$$
$$= \lim_{x \to c} \frac{x^3 - c^3}{x - c}$$
$$= \lim_{x \to c} \frac{(x - c)(x^2 + cx + c^2)}{x - c}$$
$$= \lim_{x \to c} (x^2 + cx + c^2)$$
$$= c^2 + c^2 + c^2$$

Thus, $f'(x) = 3x^2$ for all $x \in \mathbb{R}$.

(b) Let $g(x) = \frac{1}{x}$ for $x \in \mathbb{R}, x \neq 0$. For $c \in \mathbb{R}$ with $c \neq 0$, by definition of derivative, we have

$$g'(c) = \lim_{x \to c} \frac{g(x) - g(c)}{x - c} = \lim_{x \to c} \frac{\frac{1}{x} - \frac{1}{c}}{x - c} = \lim_{x \to c} \frac{\frac{c-x}{cx}}{x - c}$$

$$= \lim_{x \to c} \left[\left(-\frac{x - c}{cx} \right) \cdot \left(\frac{1}{x - c} \right) \right] = \lim_{x \to c} \left[-\frac{1}{cx} \right] = -\frac{1}{c^2}.$$

Thus, $g'(x) = -\frac{1}{x^2}$ for all $x \in \mathbb{R}$ with $x \neq 0$.

(c) Let $h := \sqrt{x}$ for $x > 0$. For $c > 0$, by definition of derivative, we have

$$h'(c) = \lim_{x \to c} \frac{h(x) - h(c)}{x - c}$$

$$= \lim_{x \to c} \frac{\sqrt{x} - \sqrt{c}}{x - c}$$

$$= \lim_{x \to c} \left(\frac{\sqrt{x} - \sqrt{c}}{x - c} \cdot \frac{\sqrt{x} + \sqrt{c}}{\sqrt{x} + \sqrt{c}} \right)$$

$$= \lim_{x \to c} \frac{x - c}{(\sqrt{x} + \sqrt{c})(x - c)}$$

$$= \lim_{x \to c} \frac{1}{\sqrt{x} + \sqrt{c}} = \frac{1}{\sqrt{c} + \sqrt{c}}$$

$$= \frac{1}{2\sqrt{c}}.$$

Thus, $h'(x) = \frac{1}{2\sqrt{x}}$ for all $x > 0$.

(d) Let $k(x) = \frac{1}{\sqrt{x}}$ for $x > 0$. For $c > 0$, by definition of derivative, we have

$$k'(c) = \lim_{x \to c} \frac{k(x) - k(c)}{x - c}$$

$$= \lim_{x \to c} \frac{\frac{1}{\sqrt{x}} - \frac{1}{\sqrt{c}}}{x - c} = \lim_{x \to c} \frac{\frac{\sqrt{c} - \sqrt{x}}{\sqrt{cx}}}{x - c}$$

$$= \lim_{x \to c} \frac{\frac{\sqrt{c} - \sqrt{x}}{\sqrt{cx}} \cdot \frac{\sqrt{c} + \sqrt{x}}{\sqrt{c} + \sqrt{x}}}{x - c}$$

$$= \lim_{x \to c} \left[\left(-\frac{x - c}{\sqrt{c^2 x} + \sqrt{cx^2}} \right) \cdot \left(\frac{1}{x - c} \right) \right]$$

$$= \lim_{x \to c} \left[-\frac{1}{\sqrt{c^2 x} + \sqrt{cx^2}} \right]$$

$$= -\frac{1}{\sqrt{c^3} + \sqrt{c^3}} = -\frac{1}{2\sqrt{c^3}}$$

$$= -\frac{1}{2} c^{-\frac{3}{2}}.$$

Therefore, $k'(x) = -\frac{1}{2} x^{-\frac{3}{2}}$ for all $x > 0$.

2. Show that $f(x) := x^{1/3}$, $x \in \mathbb{R}$, is not differentiable at $x = 0$.

Solution: $f(x) = x^{\frac{1}{3}}$.

$$\lim_{x \to 0} \frac{f(x) - f(0)}{x - 0} = \lim_{x \to 0} \frac{x^{\frac{1}{3}} - 0}{x}$$

$$= \lim_{x \to 0} \frac{x^{\frac{1}{3}}}{x^{\frac{1}{3}} \cdot x^{\frac{2}{3}}}$$

$$= \lim_{x \to 0} \frac{1}{x^{\frac{2}{3}}} \text{ not a finite quantity.}$$

Therefore, $\displaystyle\lim_{x \to 0} \frac{f(x) - f(0)}{x - 0}$ does not exist and hence f is not differentiable at $x = 0$.

3. Let $f : \mathbb{R} \to \mathbb{R}$ be defined by

$$f(x) = \begin{cases} x^2, & \text{for x rational} \\ 0, & \text{for x irrational} \end{cases}$$

Show that f is differentiable at $x = 0$, and find $f'(0)$.

Solution: Observe that 0 is a rational number and $f(0) = 0^2 = 0$.

Let

$$g(x) := \frac{f(x) - f(0)}{x - 0} = \frac{f(x)}{x}.$$

Then for $x \neq 0$, we have

$$g(x) = \begin{cases} x, & x \quad \text{rational} \\ 0, & x \quad \text{irrational} \end{cases}$$

The given function $f(x)$ is differentiable at zero if and only if $\displaystyle\lim_{x \to 0} g(x)$ exists.

In our case, the value of the limit is $f'(0)$.

We have

$$-|x| \leq g(x) \leq |x|.$$

Also,

$$\lim_{x \to 0} -|x| = \lim_{x \to 0} |x| = 0.$$

Hence, by the squeeze theorem

$$\lim_{x \to 0} g(x) = 0.$$

Therefore, f is differentiable at $x = 0$ and $f'(0) = 0$.

4. Differentiate and simplify.

(i) $f(x) = \dfrac{x}{1 + x^2}$,   (ii) $g(x) = \sqrt{5 - 2x + x^2}$,

(iii) $h(x) = (\sin x^k)^m$ for $k, k \in \mathbb{N}$ (iv) $k(x) = \tan(x^2)$ for $|x| < \sqrt{\pi/2}$.

Solution: (i) Here $f(x) = \dfrac{x}{1 + x^2}$

Let $g(x) = x$ and $h(x) = 1 + x^2$.

Therefore, $'(\ ) = 1, h'(\ ) = 2$

Now $f(x) = \dfrac{g(x)}{h(x)}$.

Therefore, by quotient formula, we get

$$
\begin{aligned}
f'(x) &= \frac{h(x)g'(x) - h'(x)g(x)}{(h(x))^2} \\
&= \frac{(1+x^2) \cdot 1 - 2x \cdot x}{(h(x))^2} \\
&= \frac{1 - x^2}{(1 + x^2)^2}.
\end{aligned}
$$

(ii) Here $g(x) = \sqrt{5 - 2x + x^2}$.

Let $h(x) = 5 - 2x + x^2$ and $f(x) = x^{\frac{1}{2}}$.

Then $g(x) = (foh)(x) = f(h(x))$.

Now $h'(x) = -2 + 2x.$ $f'(x) = \frac{1}{2}x^{-\frac{1}{2}}$.

Then by chain rule, we obtain

$$
\begin{aligned}
g'(x) &= f'(h(x))h'(x) \\
&= \frac{1}{2}h(x)^{-\frac{1}{2}}(-2 + 2x) \\
&= \frac{1}{2}(5 - 2x + x^2)^{-\frac{1}{2}}2(x - 1) \\
&= \frac{x - 1}{\sqrt{5 - 2x + x^2}}.
\end{aligned}
$$

(iii) Here $h(x) = (\sin x^k)^m$.

Let $f(x) = x^k,$ $g(x) = \sin x$ and $k(x) = x^m$.

Therefore,

$$h(x) = (kogof)(x) = k(g(f(x)))$$

. Now, $f'(x) = kx^{k-1},$ $g'(x) = \cos x$ and $h'(x) = mx^{m-1}$.

Since, $h(x) = (kogof)(x)$, then by repeated application of chain rule, we get

$$
\begin{aligned}
h'(x) &= k'(g(f(x)))g'(f(x))f'(x) \\
&= m(\sin x^k)^{m-1}\cos x^k \cdot kx^{k-1} \\
&= mkx^{k-1}\cos x^k(\sin x^k)^{m-1}.
\end{aligned}
$$

(iv) Here $k(x) = \tan x^2$.

Let $f(x) = x^2$ and $g(x) = \tan x$.

Therefore, $f'(x) = 2x,$ $g'(x) = \sec^2 x$

Now, $k(x) = (fog)(x)$, then by chain rule, we have

$$
\begin{aligned}
k'(x) &= q'(p(x))p'(x) \\
&= \sec^2(x^2)2x
\end{aligned}
$$

5. Let $n \in \mathbb{N}$ and let $f : \mathbb{R} \to \mathbb{R}$ be defined by

$$f(x) = \begin{cases} x^n, & for \ x \geq 0, \\ 0, & for \ x < 0. \end{cases}$$

For which values of n is f' continuous at 0? For which values of n is f' differentiable at 0?

Solution: Let $n \in \mathbb{N}$ and let $f : \mathbb{R} \to \mathbb{R}$ be defined by $f(x) = \begin{cases} x^n, & for \ x \geq 0, \\ 0, & for \ x < 0. \end{cases}$

Clearly, $f'(x) = \begin{cases} nx^{n-1}, & for \ x > 0, \\ 0, & for \ x < 0. \end{cases}$

We have to obtain $f'(0)$.

$$\begin{aligned} f'_-(0) &= \lim_{x \to 0^-} \frac{f(x) - f(0)}{x - 0} \\ &= 0 \qquad\qquad (since, \ f(x) = 0 \ if \ x < 0). \\ f'_+(0) &= \lim_{x \to 0^+} \frac{f(x) - f(0)}{x - 0} \\ &= \lim_{x \to 0} \frac{x^n}{x} \\ &= \begin{cases} 1, & if \ n = 1, \\ 0, & if \ n > 1. \end{cases} \end{aligned}$$

Therefore, we conclude that

$$f'_-(0) = f'_+(0) = 0 \ \ if \ \ n > 1.$$

$$\text{But } f'(0^-) \neq f'(0^+) \ \ if \ \ n = 1.$$

Hence we conclude that

$$f'(x) = \begin{cases} nx^{n-1}, & x > 0 \ \ if \ \ n > 1, \\ 0, & x \leq 0. \end{cases}$$

and $f'(0)$ does not exist if $n = 1$.

Therefore, f' is continuous at $x = 0$ for $n > 1$ and differentiable at $x = 0$ for $n > 1$.

6. Suppose that $f : \mathbb{R} \to \mathbb{R}$ is differentiable at c and that $f(c) = 0$. Show that $g(x) = |f(x)|$ is differentiable at c if and only if $f'(c) = 0$.

Solution: Suppose, f is differentiable at c and $f(c) = 0$.

Let $g = |f|$ be differentiable at c. Therefore, the following limit exists.

$$\begin{aligned} g'(c) &= \lim_{x \to c} \frac{g(x) - g(c)}{x - c} \\ &= \lim_{x \to c} \frac{|f(x)| - |f(c)|}{x - c} \\ &= \lim_{x \to c} \frac{|f(x)|}{x - c} \qquad (f(c) = 0) \end{aligned}$$

Suppose $f'(c) = L \neq 0$. Then for $x_n = c \pm \frac{1}{n}$, we get

$$\lim \ \frac{f(x_n)}{} \qquad L \iff \lim \ \frac{f(c \pm \frac{1}{n})}{} \qquad L$$

Then

$$\lim_{n \to \infty} \frac{|f(c \pm \frac{1}{n})|}{\pm \frac{1}{n}} = \pm L$$

Therefore, $\lim_{x \to c} \frac{|f(x)|}{x - c}$ does not exists.

Which is contradiction to $g(x) = |f(x)|$ is differentiable at $x = 0$.

Therefore, $f'(c) = 0$. Conversely, if $f'(c) = 0$, then

$$0 = |f'(c)| = \left|\lim_{x \to c} \frac{f(x)}{x - c}\right| = \lim_{x \to c} \left|\frac{f(x)}{x - c}\right| = \lim_{x \to c} \frac{|f(x)|}{|x - c|}$$

Let $\epsilon > 0$ arbitrary. There exists $\delta > 0$ such that

$$|x - c| < \delta \Rightarrow \frac{|f(x)|}{|x - c|} < \epsilon$$

Now, if $|x - c| < \delta \Rightarrow$

$$\frac{|f(x)|}{x - c} \leq \left|\frac{|f(x)|}{x - c}\right| = \frac{|f(x)|}{|x - c|} < \epsilon$$

Hence, $g'(c)$ exists and $g'(c) = 0$.

Therefore, $g(x) = |f(x)|$ is differentiable at $x = 0$.

7. Determine where each of the following functions from $\mathbb{R}$ to $\mathbb{R}$ is differentiable and find the derivative.

(a) $f(x) = |x| + |x + 1|$,   (b) $g(x) = 2x + |x|$,

(c) $h(x) = x|x|$,   (d) $k(x) = |\sin x|$.

Solution: (a) Here $f(x) := |x| + |x + 1|$.

$$f(x) = \begin{cases} 2x + 1, & x \geq 0 \\ 1, & -1 \leq x < 0 \\ -2x - 1, & x < -1. \end{cases}$$

For $x \geq 0$, $f(x) = 2x + 1 \Rightarrow f'(x) = 2$ and for $x \in [-1, 0)$, $f(x) = 1 \Rightarrow f'(x) = 0$

For $x < -1$, $f(x) = -2x - 1 \Rightarrow f'(x) = -2$

It remains to check differentiable at point $x = -1$ and $x = 0$.

$$\lim_{x \to 0^+} \frac{f(x) - f(0)}{x - 0} = \lim_{x \to 0^+} \frac{2x + 1 - (2 \cdot 0 + 1)}{x} = 2$$

$$\lim_{x \to 0^-} \frac{f(x) - f(0)}{x - 0} = \lim_{x \to 0^-} \frac{1 - 1}{x} = 0$$

Therefore, $f'(0)$ does not exists.

$$\lim_{x \to -1^+} \frac{f(x) - f(-1)}{x + 1} = \lim_{x \to -1^+} \frac{1 - 1}{x + 1} = 0$$

$$\lim_{1} \frac{f(x) - f(-1)}{x + 1} = \lim_{1} \frac{-2x - 1 - (2 \cdot (-1) - 1)}{x + 1} = -2.$$

Therefore, $f'(-1)$ also does not exist.

Hence, we conclude that

$$f'(x) = \begin{cases} 2, & x < 0 \\ 0, & -1 < x < 0 \\ -2, & x < -1 \\ for\ x\ = -1, x = 0 \text{ does not exists.} \end{cases}$$

(b) Here, $g(x) = 2x + |x|$.

$$g(x) = \begin{cases} 3x, & x \geq 0 \\ x, & x < 0 \end{cases}$$

For $x > 0$ $g(x) = 3x \Rightarrow g'(x) = 3$ and for $x < 0$ $g(x) = x \Rightarrow g'(x) = 1$.

$$\lim_{x \to 0^-} \frac{g(x) - g(0)}{x - 0} = \lim_{x \to 0^-} \frac{x}{x} = 1,$$

$$\lim_{x \to 0^+} \frac{g(x) - g(0)}{x - 0} = \lim_{x \to 0^+} \frac{3x}{x} = 3.$$

Therefore, g is not differentiable at $x = 0$.

(c) Here $h(x) = x|x|$.

$$h(x) = \begin{cases} x^2, & x \geq 0 \\ -x^2, & x < 0 \end{cases}$$

For $x < 0$ $g(x) = -x^2 \Rightarrow g'(x) = -2x$

For $x > 0$ $g(x) = x^2 \Rightarrow g'(x) = 2x$

$$\lim_{x \to 0^+} \frac{h(x) - h(0)}{x - 0} = \lim_{x \to 0^+} \frac{x^2}{x} = 0$$

$$\lim_{x \to 0^-} \frac{h(x) - h(0)}{x - 0} = \lim_{x \to 0^-} \frac{-x^2}{x} = 0$$

Hence, h is differentiable at $x = 0$.

Therefore, $h'(0)$ exists and $h'(0) = 0$.

Therefore, we conclude that,

$$h'(x) = \begin{cases} 2x, & x \geq 0 \\ -2x, & x < 0 \end{cases}$$

$$h'(x) = 2|x|.$$

(d) Here $k(x) = |\sin x|$.

For $x \in (n\pi, (n+1)\pi)$, $\sin x > 0$ if n is even and $\sin x < 0$ if n is odd.

Also, $\sin(n\pi) = 0$, $\forall\ n \in \mathbb{Z}$.

Therefore,

$$k(x) = |\sin x| = \begin{cases} \sin x, & x \in \langle 2l\pi, (2l+1)\pi\rangle, & l \in \mathbb{Z} \\ -\sin x, & x \in \langle(2l-1)\pi, 2l\pi\rangle, & l \in \mathbb{Z} \\ 0, & x = l\pi, & l \in \mathbb{Z} \end{cases}$$

Now, we have

$$k'(x) = \begin{cases} \cos x, & x \in \langle 2l\pi, (2l+1)\pi\rangle, \quad l \in \mathbb{Z} \\ -\cos x, & x \in \langle (2l-1)\pi, 2l\pi\rangle, \quad l \in \mathbb{Z} \end{cases}$$

To show that $k(x)$ is not differentiable at points $x = l\pi, \ l \in \mathbb{Z}$.

For that we define $x_n = l\pi + \frac{1}{n}, n \in \mathbb{N}$.

$$\lim_{n\to\infty} \frac{|k(x_n)| - |k(l\pi)|}{x_n - l\pi} = \lim_{n\to\infty} \frac{|\sin x_n| - |\sin l\pi|}{x_n - l\pi}.$$

$$= \lim_{n\to\infty} \frac{|\sin(l\pi + \frac{1}{n})|}{l\pi + \frac{1}{n} - l\pi} = \lim_{n\to\infty} \frac{|\sin(l\pi + \frac{1}{n})|}{l\pi + \frac{1}{n} - l\pi}$$

$$= \lim_{n\to\infty} \frac{|\sin \frac{1}{n}|}{\frac{1}{n}} \quad (\sin(l\pi + c) - \sin c, \ \forall \, l \in \mathbb{Z}, \ \forall \, c \in \mathbb{R})$$

$$= \lim_{n\to\infty} \frac{\sin \frac{1}{n}}{\frac{1}{n}} \quad \left(\sin \frac{1}{n} > 0, \ \forall \, n \in \mathbb{N}\right)$$

$$= 1 \qquad \left(\lim_{x\to 0} \frac{\sin x}{x} = 1\right)$$

Therefore, $k(x_n)$ does not converge to $k(l\pi) = 0$.

Hence, $k(x)$ is not differentiable at points $x = l\pi, \ l \in \mathbb{Z}$.

8. Prove that if $f : \mathbb{R} \to \mathbb{R}$ is an even function (that is, $f(-x) = f(x)$ for all $x \in \mathbb{R}$) and has a derivative at every point, then the derivative f' is an odd function (that is, $f'(-x) = -f'(x)$ for all $x \in \mathbb{R}$). Also prove that if $g : \mathbb{R} \to \mathbb{R}$ is a differentiable odd function, then g' is an even function.

Solution: Suppose $g(x)$ is an odd function $\implies g(-x) = -g(x), \ \forall \, x$

$$\begin{aligned} g'(c) &= \lim_{x\to c} \frac{g(x) - g(c)}{x - c} \\ g'(-c) &= \lim_{x\to -c} \frac{g(x) - g(-c)}{x - (-c)} \\ &\quad \text{change every } x \text{ for } -x \text{ if } -x \to -c \text{ then } x \to c \\ &= \lim_{x\to c} \frac{g(-x) - (-g(c))}{-x + c} \quad g \text{ is an odd function} \\ &= \lim_{x\to c} \frac{-g(x) + g(c)}{-x + c} = \lim_{x\to c} \frac{-(g(x) - g(c))}{-(x - c)} \\ &= \lim_{x\to c} \frac{g(x) - g(c)}{x - c} \\ \therefore g'(-c) &= g'(c) \end{aligned}$$

Therefore, g' is an even function.

Now, f is an even function $\implies f(-x) = f(x), \quad \forall\, x$

$$
\begin{aligned}
f'(c) &= \lim_{x \to c} \frac{f(x) - f(c)}{x - c} \\
f'(-c) &= \lim_{x \to -c} \frac{f(x) - f(-c)}{x - (-c)} \\
&\quad \text{change every } x \text{ for } -x \text{ if } -x \to -c \text{ then } x \to c \\
&= \lim_{x \to c} \frac{f(-x) - f(c)}{-x + c} \quad f \text{ is an odd function} \\
&= \lim_{x \to c} \frac{f(x) - f(c)}{-(x - c)} = \lim_{x \to c} \frac{f(x) - f(c)}{x - c} \\
&= f'(c)
\end{aligned}
$$

$f'(-c) = f'(c)$ and therefore f' is an even function.

9. Let $g : \mathbb{R} \to \mathbb{R}$ be defined by

$$
g(x) = \begin{cases} x^2 \sin(1/x^2), & for\ x \neq 0, \\ 0, & for\ x = 0 \end{cases}
$$

Show that g is differentiable for all $x \in \mathbb{R}$. Also show that the derivative g' is not bounded on the interval $[-1, 1]$.

Solution: Here

$$
g(x) = \begin{cases} x^2 \sin \frac{1}{x^2} & x \neq 0 \\ 0 & x = 0 \end{cases}
$$

To obtain the derivation of g for $x = 0$.

$$
\begin{aligned}
\lim_{x \to 0} \frac{g(x) - g(0)}{x - 0} &= \lim_{x \to 0} \frac{x^2 \sin \frac{1}{x^2}}{x} \\
&= \lim_{x \to 0} x \sin \frac{1}{x} \\
&\qquad\qquad \lim_{x \to 0} x = 0 \text{ and } \sin\left(\frac{1}{x}\right) \text{ is bounded} \\
&= 0
\end{aligned}
$$

Therefore,

$$
\lim_{x \to 0} \frac{g(x) - g(0)}{x - 0} = 0 \implies g'(0) = 0.
$$

For $x \neq 0$, we have

$$
\begin{aligned}
g(x) &= x^2 \sin \frac{1}{x^2} \\
g'(x) &= \frac{d}{dx}(x^2) \sin \frac{1}{x^2} + x^2 \frac{d}{dx}\left(\sin \frac{1}{x^2}\right) \quad \text{(by product rule} \\
&= 2x \sin \frac{1}{x^2} + x^2 \cos \frac{1}{x^2} \frac{d}{dx}\left(\frac{1}{x^2}\right) \quad \text{(chain rule} \\
&= 2x \sin \frac{1}{x^2} + x^2 \cos \frac{1}{x^2} \left(\frac{-2}{x^3}\right) \\
&\qquad\quad 1 \qquad 2 \qquad 1
\end{aligned}
$$

Therefore, function $g(x)$ is differentiable and its derivative is

$$g'(x) = \begin{cases} 2\left(x\sin\frac{1}{x^2} - \frac{1}{x}\cos\frac{1}{x^2}\right), & \text{for} \quad x \neq 0 \\ 0, & \text{for} \quad x = 0 \end{cases}$$

10. Assume that there exists a function $L : (0, \infty) \to \mathbb{R}$ such that $L'(x) = 1/x$ for $x > 0$. Calculate the derivatives of the following functions

(a) $f(x) = L(2x + 3)$ for $x > 0$,   (b) $g(x) = (L(x^2))^3$ for $x > 0$,

(c) $h(x) = L(ax)$ for $a > 0, x > 0$,   (d) $k(x) = L(L(x))$ when $L(x) > 0, x > 0$.

Solution: Suppose $L : (0, \infty) \to \mathbb{R}$ is a function such that $L'(x) = 1/x$ for $x > 0$.

(a) Here $f(x) = L(2x + 3)$ for $x > 0$.

$$\begin{aligned} f'(x) &= \frac{d}{dx}(L(2x + 3)) && \text{(by Chain Rule)} \\ &= \frac{1}{2x + 3} \cdot \frac{d}{dx}(2x + 3) \\ &= \frac{2}{2x + 3} \end{aligned}$$

(b) Here $g(x) = (L(x^2))^3$ for $x > 0$.

$$\begin{aligned} g'(x) &= \frac{d}{dx}((L(x^2))^3) && \text{(by Chain Rule)} \\ &= 3(L(x^2))^2 \cdot \frac{d}{dx}(L(x^2)) && \text{(by Chain Rule)} \\ &= 3(L(x^2))^2 \cdot \frac{1}{x^2} \cdot \frac{d}{dx}(x^2) \\ &= 3(L(x^2))^2 \cdot \frac{1}{x^2} \cdot 2x \\ &= \frac{6 \cdot (L(x^2))^2}{x} \end{aligned}$$

(c) Here $h(x) = L(ax)$ for $a > 0, x > 0$.

$$\begin{aligned} h'(x) &= \frac{d}{dx}(L(ax)) && \text{(by Chain Rule)} \\ &= \frac{1}{ax} \cdot \frac{d}{dx}(ax) \\ &= \frac{1}{dx} \cdot a \\ &= \frac{1}{x} \end{aligned}$$

(d) Here $k(x) = L(L(x))$ when $L(x) > 0, x > 0$.

$$\begin{aligned} k'(x) &= \frac{d}{dx}(L(L(x))) && \text{(by Chain Rule)} \\ &= \frac{1}{L(x)} \cdot \frac{d}{dx}(L(x)) \\ &= \frac{1}{L(x)} \cdot \frac{1}{x} \\ &= \frac{1}{} \end{aligned}$$

11. If $r > 0$ is a rational number, let $f : \mathbb{R} \to \mathbb{R}$ be defined by

$$f(x) = \begin{cases} x^r \sin(1/x), & for\ x \neq 0, \\ 0, & for\ x = 0 \end{cases}$$

Determine those values of r for which $f'(0)$ exists.

Solution: Here

$$f(x) = \begin{cases} x^r \sin(1/x), & for\ x \neq 0, \\ 0, & for\ x = 0 \end{cases}$$

Now, the function f is piecewise defined function and we have a problem at $x = 0$. To obtain the limit of $f(x)$ as $x \to 0$.

$$\lim_{x \to 0} \frac{f(x) - f(0)}{x - 0} = \lim_{x \to 0} \frac{x^r \sin \frac{1}{x}}{x}$$
$$= \lim_{x \to 0} x^{r-1} \sin\left(\frac{1}{x}\right).$$

Case (i): For $r > 1$

We know that $\lim\limits_{x \to 0} x = 0$ and that $sin(\frac{1}{x})$ is bounded.

$$\therefore \lim_{x \to 0} x^{r-1} \sin\left(\frac{1}{x}\right) = 0.$$

Case (ii): For $r = 1$

Therefore, $\lim\limits_{x \to 0} sin\left(\frac{1}{x}\right)$ does not exists.

Case (iii): For $r < 1$

In this case $\dfrac{1}{x^{r-1}}$ is not bounded in the neighborhood of 0.

$\therefore \lim\limits_{x \to 0} x^{r-1} \sin\left(\frac{1}{x}\right)$ does not exist.

Hence, we conclude that $f'(0)$ only exists when $r > 1$.

1.6   The Mean Value Theorem

In Calculus there is the important result that, the Mean Value Theorem, which relates the values of a function to values of its derivative. In this section, we will establish this important theorem and their applications. Now, first we see the relationship between the relative extrema of a function and the values of its derivative.

Definition 1.3. *A (real valued) function $f : I \to \mathbb{R}$ is said to have*

(a) **global maximum** *at a point $c \in II$ if $f(x) \leq f(c)$ for all $x \in I$, and*

(b) **global minimum** *at a point $c \in II$ if $f(x) \geq f(c)$ for all $x \in I$.*

We say that f has a **global extremum** at $c \in I$ if it has either a relative maximum or a relative minimum at c.

Definition 1.4. *A (real valued) function $f : I \to \mathbb{R}$ is said to have*

(a) **relative or local maximum** *at a point $c \in I$ if there exist $\delta > 0$ such that*

$$f(x) \le f(c), \ \forall\, x \in I \cap (c - \delta, c + \delta)$$

(b) **relative or local minimum** *at a point $c \in I$ if there exist $\delta > 0$ such that*

$$f(x) \ge f(c), \ \forall\, x \in I \cap (c - \delta, c + \delta).$$

The function f is said to attain **relative or local extremum** at a point $c \in I$ if f attains either local maximum or local minimum at c.

Remark 1.3. *It is conventional to omit the adjective local in local maximum, local minimum and local extremum. Thus when we say a function has maximum at a point c, we generally mean a local maximum at c. Similar comments apply to minimum and extremum.*

The next result provides the theoretical justification for the familiar process of finding points at which f has relative extrema by examining the zeros of the derivative. However, it must be realized that this procedure applies only to interior points of the interval.

For example, if $f(x) = x$ on the interval $I = [0, 1]$, then the endpoint $x = 0$ yields the unique relative minimum and the endpoint $x = 1$ yields the unique maximum of f on I, but neither point is a zero of the derivative of f.

Theorem 1.7. *Interior Extremum Theorem*

Let c be an interior point of the interval I at which $f : I \to \mathbb{R}$ has a relative extremum. If the derivative of f at c exists, then $f'(c) = 0$.

Proof. We will prove the result only for the case that f has a relative maximum at c; the proof for the case of a relative minimum is similar.

Suppose f have a relative maximum at c which is an interior point of I.

Then there exists $\delta > 0$ such that $(c - \delta, c + \delta) \subseteq I$ and $f(c) \ge f(c + h)$ for all h with $|h| < \delta$.

Hence, for all h with $|h| < \delta$, we have

$$\frac{f(c + h) - f(c)}{h} \ge 0 \text{ if } h < 0,$$

$$\frac{f(c + h) - f(c)}{h} \le 0 \text{ if } h > 0.$$

Taking limit as $h \to 0$, we get

$$f'(c) = \lim_{h \to 0} \frac{f(c + h) - f(c)}{h} \ge 0,$$

$$f'(c) = \lim_{\to 0} \frac{f(c + h) - f(c)}{h} \le 0.$$

Corollary 1.2. *Let $f : I \to \mathbb{R}$ be continuous on an interval I and suppose that f has a relative extremum at an interior point c of I. Then either the derivative of f at c does not exist, or it is equal to zero.*

Note: (i) if $f(x) = |x|$ on $I := [-1, 1]$, then f has an interior minimum at $x = 0$. however, the derivative of f fails to exist at $x = 0$.

(ii) A function can have more than one maximum and minimum.

For example, consider $f(x) = sin(4x)$, $x \in [0, \pi]$.

We see that f has maximum value 1 at $\pi/8$ and $5\pi/8$, and has minimum value -1 at $3\pi/8$ and $7\pi/8$.

(iii) If a function f is differentiable at an interior point c of an interval I and $f'(c) \neq 0$, then f can not have local maximum or local minimum at c.

(iv) A function f have a maximum or minimum at a point c, the function need not be differentiable at c.

For example

$$f(x) = 1 - |x|, \ |x| \leq 1,$$

has a maximum at 0 and

$$g(x) = |x|, |x| \leq 1,$$

has a minimum at 0. Both f and g are not differentiable at 0.

Theorem 1.8. *Rolles Theorem*

Suppose that f is continuous on a closed interval $I := [a, b]$, that the derivative f' exists at every point of the open interval (a, b), and that $f(a) = f(b) = 0$. Then there exists at least one point c in (a, b) such that $f'(c) = 0$.

Proof. Let $g(x) = f(x) - f(a)$.

Therefore, we have $g(a) = 0 = g(b)$, and $g'(x) = f'(x)$ for every $x \in (a, b)$.

Now, g have a maximum and minimum at some points x_1 and x_2, respectively in $[a, b]$

That is there exists $x_1, \ x_2 \in [a, b]$ such that

$$g(x_2) \leq g(x) \leq g(x_1), \ \forall \ x \in [a, b].$$

Case(i): If $g(x_1) = g(x_2)$, then g is a constant function and hence $g'(x) = 0, \ \forall \ x \in [a, b]$.

Case(ii): We assume that $g(x_2) < g(x_1)$.

Then, either $g(x_1) \neq 0$ or $g(x_2) \neq 0$.

Suppose that $g(x_2) \neq 0$, so that $x_2 \neq a$ and $x_2 \neq b$.

Hence, by Interior Extremum Theorem, we get

$$g'(x_2) = 0.$$

Therefore, $f'(x_2) = 0$.

Case(iii): Similarly, if $g(x_1) \neq 0$, then we will prove $f'(x_1) = 0$.

Hence, theorem is proved.

Remark 1.4. *Geometrical Interpretation of Rolle's theorem:*

Suppose that f is continuous on a closed interval $I := [a, b]$, that the derivative f' exists at every point of the open interval (a, b) and that $f(a) = f(b) = 0$. Then there is a point c on the interval (a, b) where the tangent to the graph of the function is parallel to x-axis.

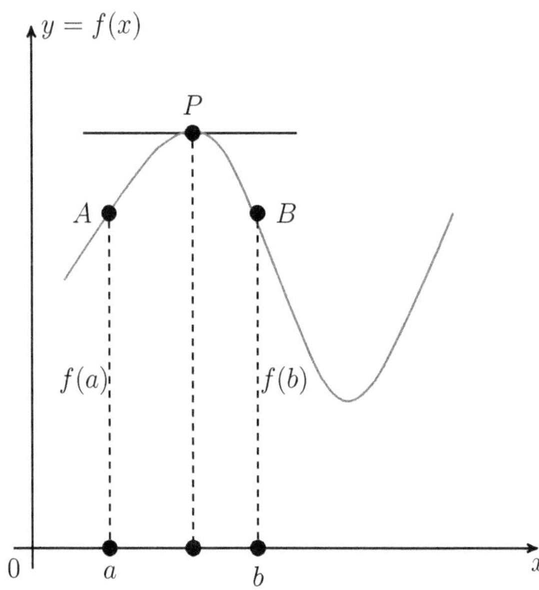

Example 1.9. *Consider $f(x) = |x|$ on closed interval $[-1, 1]$. This function is continuous on closed interval $[-1, 1]$ but this function does not have derivative at $x = 0$. Therefore, Rolle's theorem fails here because $f(x)$ is not differentiable over the whole interval $(-1, 1)$.*

Example 1.10. *Consider $f(x) = x$ on closed interval $[0, 1]$. This function is continuous on closed interval $[0, 1]$ and differentiable on open interval $(0, 1)$ but $f(0) \neq f(1)$. Therefore, the condition of Rolle's theorem is $f(0) = f(1)$ is not satisfied, hence Rolle'e theorem fails here.*

Example 1.11. *Let $f(x) = x^2 + 2x$. Find all values of c in the interval $[-2, 0]$ such that $f'(c) = 0$.*

Solution: We check that the function $f(x)$ satisfies all conditions of Rolle's theorem.

Now, $f(x) = x^2 + 2x$ is quadratic function therefore, it is continuous in $[-2, 0]$ and it is differentiable over the open interval $(-2, 0)$.

Also,

$$f(-2) = (-2)^2 + 2.(-2) = 0 = f(0)$$

$$f'(x) = 2x + 2.$$

Therefore, f satisfies all conditions of Rolle's theorem.

Hence, by Rolle's theorem there exists $c \in (-2, 0)$ such that

$$f'(c) = 0.$$

$$\therefore 2c + 2 = 0 \Rightarrow c = -1.$$

Example 1.12. *Verify the Rolle's theorem for the function $f(x) = x^2 - 6x + 8$ on $[2, 4]$, if so find the value of c.*

Solution: Let $f(x) = x^2 - 6x + 8$ on $[2, 4]$.

Now, $f(x) = x^2 - 6x + 8$ is quadratic function therefore, it is continuous in $[2, 4]$ and it is differentiable over the open interval $(2, 4)$.

Also,

$$f(2) = (2)^2 - 6.2 + 8 = 0$$

$$f(4) = (4)^2 - 6.4 + 8 = 0$$

$$\therefore f(2) = f(4).$$

Therefore, f satisfies all conditions of Rolle's theorem.

Therefore, by Rolle's theorem there exists $c \in (2, 4)$ such that

$$f'(c) = 0.$$

$$\therefore 2c - 6 = 0 \Rightarrow c = 3.$$

$$\therefore c = 3 \in (2, 4).$$

Example 1.13. *Verify the Rolle's theorem for the function $f(x) = \sqrt{1 - x^2}$ on $[-1, 1]$, if so find the value of c.*

Solution: Let $f(x) = \sqrt{1 - x^2}$ on $[-1, 1]$.

Now, $\phi(x) = 1 - x^2$ is continuous in $[-1, 1]$ and it is differentiable over the open interval $(-1, 1)$.

Therefore, $f(x) = \sqrt{\phi(x)}$ is continuous in $[-1, 1]$ and it is differentiable over the open interval $(-1, 1)$.

Also,

$$f(-1) = 8 = f(1)$$

Therefore, f satisfies all conditions of Rolle's theorem.

Therefore, by Rolle's theorem there exists $c \in (2, 4)$ such that

$$f'(c) = 0.$$

$$f(x) = \sqrt{1 - x^2}$$
$$\Rightarrow f'(x) = -\frac{x}{\sqrt{1 - x^2}}$$
$$\therefore f'(c) = 0$$
$$\Rightarrow -\frac{c}{\sqrt{1 - c^2}} = 0 \Rightarrow c = 0.$$
$$\therefore c = 0 \in (-1, 1).$$

Theorem 1.9. *Mean Value Theorem*

Suppose that f is continuous on a closed interval $I := [a, b]$, and that f has a derivative in the open interval (a, b). Then there exists at least one point c in (a, b) such that

$$f(b) - f(a) = f'(c)(b - a).$$

Proof. Consider the function ϕ defined on I by

$$\phi(x) = f(x) - f(a) - \frac{f(b) - f(a)}{b - a}(x - a).$$

The function ϕ is simply the difference of f and the function whose graph is the line segment joining the points $(a, f(a))$ and $(b, f(b))$.

Clearly, ϕ is continuous on $[a, b]$, differentiable on (a, b), and $\phi(a) = \phi(b) = 0$.

Now, $\phi(x)$ satisfies all conditions of Rolles theorem. Therefore, by Rolles theorem there exists a point c in (a, b) such that

$$\phi'(c) = 0.$$
$$\therefore f'(c) - \frac{f(b) - f(a)}{b - a} = 0$$

Hence,

$$f(b) - f(a) = f'(c)(b - a)$$

is proved.

Remark 1.5. *The geometrical meaning of the Mean Value Theorem is that there is some point on the curve $y = f(x)$ at which the tangent line is parallel to the line segment through the points $(a, f(a))$ and $(b, f(b))$.*

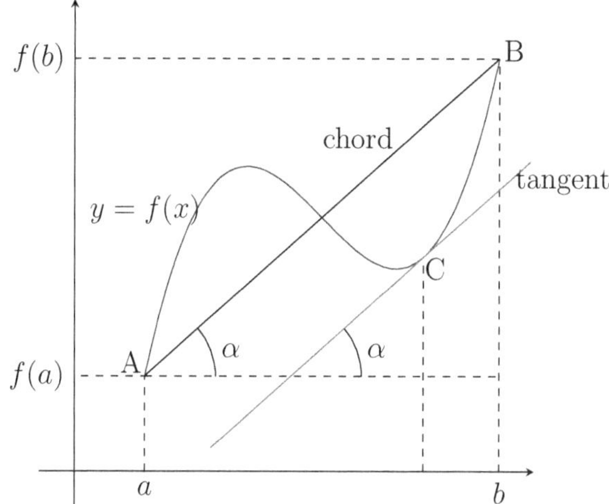

The Mean Value Theorem permits one to draw conclusions about the nature of a function f from information about its derivative f'. The following results are obtained in this manner.

Theorem 1.10. *If a function f is continuous on the closed interval $I := [a, b]$, that f is differentiable on the open interval (a, b), and that $f'(x) = 0$ for $x \in (a, b)$ then f is constant on I.*

Proof. We will show that $f(x) = f(a)$ for all $x \in I$.

Indeed, if $x \in I, x > a$, is given, we apply the Mean Value Theorem to f on the closed interval $[a, x]$. We obtain a point c (depending on x) between a and x such that

$$f(x) - f(a) = f'(c)(x - a).$$

Since $f'(c) = 0$ (by hypothesis), we deduce that

$$f(x) - f(a) = 0.$$

Hence, $f(x) = f(a)$ for any $x \in I$. Therefore, f is constant on I.

Corollary 1.3. *Suppose that f and g are continuous on $I := [a, b]$. They are differentiable on (a, b), and $f'(x) = g'(x)$ for all $x \in (a, b)$.*
Then there exists a constant c such that $f = g + c$ on I.

Example 1.14. *Verify the Mean Value theorem for the function $f(x) = x^2 - 3x + 5$ on $[1, 4]$, if so find the value of c.*

Solution: Let $f(x) = x^2 - 3x + 5$ on $[1, 4]$.

Now, $f(x) = x^2 - 3x + 5$ is quadratic function therefore, it is continuous in $[1, 4]$ and it is differentiable on open interval $(1, 4)$.

$$f'(x) = 2x - 3$$

Therefore, f satisfies all conditions of Mean Value theorem. Hence, by Mean Value theorem there exists $c \in (1, 4)$ such that

$$f'(c) = \frac{f(b) - f(a)}{b - a}$$
$$2c - 3 = \frac{f(4) - f(1)}{4 - 1}$$
$$2c - 3 = \frac{9 - 3}{3}$$
$$2c - 3 = 2 \Rightarrow 2c = 5 \Rightarrow= \frac{5}{2} \in (1, 4).$$

Example 1.15. *Verify the Mean Value theorem for the function* $f(x) = \sqrt{x + 4}$ *on* $[0, 5]$, *if so find the value of* c.

Solution: Let $f(x) = \sqrt{x + 4}$ on $[0, 5]$.

Now, $f(x) = \sqrt{x + 4}$ is continuous in $[0, 5]$ and it is differentiable on open interval $(0, 5)$.

$$f(x) = \sqrt{x + 4}$$
$$f'(x) = \frac{1}{2\sqrt{x + 4}}$$

Therefore, f satisfies all conditions of Mean Value theorem. Hence, by Mean Value theorem there exists $c \in (0, 5)$ such that

$$f'(c) = \frac{f(b) - f(a)}{b - a}$$
$$\frac{1}{2\sqrt{c + 4}} = \frac{f(5) - f(0)}{5 - 0}$$
$$\frac{1}{2\sqrt{c + 4}} = \frac{1}{5}$$
$$\sqrt{c + 4} = \frac{5}{2}$$
$$c = \frac{9}{4} \in (0, 5).$$

Example 1.16. *Use Mean Value theorem to prove that for any two real numbers* a *and* b, $|cosa - cosb| \leq |a - b|$.

Solution: Let $f(x) = cosx$ on $[a, b]$, where a and b are real numbers.

Now, $f(x) = cosx$ is continuous in $[a, b]$ and it is differentiable on open interval (a, b).

$$f(x) = \cos x, \quad f'(x) = -\sin x$$

Therefore, f satisfies all conditions of Mean Value theorem. Hence, by Mean Value theorem there exists $c \in (a, b)$ such that

$$f'(c) = \frac{f(b) - f(a)}{b - a}$$

$$\sin c = \frac{\cos b - \cos a}{b - a}$$

$$\frac{\cos b - \cos a}{b - a} = \sin c \leq 1 \Rightarrow \frac{\cos b - \cos a}{b - a} \leq 1$$

$$\left| \frac{\cos b - \cos a}{b - a} \right| \leq 1$$

$$|\cos a - \cos b| \leq |a - b|.$$

Theorem 1.11. *Cauchy's Mean Value Theorem*

Let f and g are continuous functions on a closed interval $I = [a, b]$ and differentiable on the open interval (a, b). Suppose $g'(x) \neq 0$, $\forall x \in (a, b)$. Then there exists at least one point c in (a, b) such that

$$\frac{f(b) - f(a)}{g(b) - g(a)} = \frac{f'(c)}{g'(c)}.$$

Proof. Consider the function ϕ defined on I by

$$\phi(x) = f(x) - f(a) - \frac{f(b) - f(a)}{g(b) - g(a)}(g(x) - g(a)).$$

Clearly, ϕ is continuous on $[a, b]$, differentiable on (a, b), and $\phi(a) = \phi(b) = 0$.

Now, $\phi(x)$ satisfies all conditions of Rolles theorem. Therefore, by Rolles theorem there exists a point c in (a, b) such that

$$\phi'(c) = 0.$$

$$\therefore f'(c) - \frac{f(b) - f(a)}{g(b) - g(a)}g'(c) = 0$$

Since $g'(c) \neq 0$, therefore,

$$\frac{f(b) - f(a)}{g(b) - g(a)} = \frac{f'(c)}{g'(c)}$$

is proved.

Example 1.17. *Verify the Cauchy's Mean Value theorem for the functions $f(x) = x^2$, $g(x) = x^3$ on $[1, 2]$, if so find the value of c.*

Solution: Let $f(x) = x^2$, $g(x) = x^3$ on $[1, 2]$.

Now, f and g are polynomials in x, therefore, both are continuous in $[1, 2]$ and differentiable on open interval $(1, 2)$.

$$f'(x) = 2x, \quad g'(x) = 3x^2.$$

Now, $g'(x) \neq 0$, $\forall x \in (1, 2)$.

Therefore, f and g satisfies all conditions of Cauchy's Mean Value theorem. Hence, by Cauchy's

Mean Value theorem there exists $c \in (1, 2)$ such that

$$\frac{f(2) - f(1)}{g(2) - g(1)} = \frac{f'(c)}{g'(c)}$$

$$\frac{4 - 1}{8 - 1} = \frac{2c}{3c^2}$$

$$\frac{3}{7} = \frac{2}{3c}$$

$$2c - 3 = \frac{9 - 3}{3}$$

$$9c = 14$$

$$c = \frac{14}{9} \in (1, 2).$$

Example 1.18. *Using Cauch's Mean Value theorem show that $1 - \dfrac{x^2}{2!} < cosx$ for $x \neq 0$.*

Solution: Let $f(x) = 1 - cosx$ on $[a, b]$, $g(x) = \dfrac{x^2}{2!}$.

Now, $f(x) = 1 - cosx$ and $g(x) = \frac{x^2}{2!}$ are continuous in $[0, x]$ and differentiable on open interval $(0, x)$.

$$f'(x) = sinx, \quad g'(x) = x$$

Therefore, f and g satisfies all conditions of Cauchy's Mean Value theorem. Hence, by Cauchy's Mean Value theorem there exists $c \in (0, x)$ such that

$$\frac{f(x) - f(0)}{g(x) - g(0)} = \frac{f'(c)}{g'(c)}$$

$$\frac{1 - cosx}{\dfrac{x^2}{2!}} = \frac{sinc}{c}$$

$$\frac{1 - cosx}{\dfrac{x^2}{2!}} = \frac{sinc}{c} < 1.$$

Therefore, this proves

$$1 - \frac{x^2}{2!} < cosx, \quad \text{for} \quad x \neq 0.$$

1.7   Increasing and decreasing Function

Definition 1.5. *(i) A function $f : I \to \mathbb{R}$ is said to be **increasing** on the interval I if whenever x_1, x_2 in I satisfy $x_1 < x_2$, then $f(x_1) \leq f(x_2)$.*

*(ii) A function $f : I \to \mathbb{R}$ is said to be **decreasing** on the interval I if whenever x_1, x_2 in I satisfy $x_1 < x_2$, then $f(x_1) \geq f(x_2)$.*

Theorem 1.12. *Let $f : I \to \mathbb{R}$ be differentiable on the interval I. Then*

(a) f is increasing on I if and only if $f'(x) \geq 0$ for all $x \in I$.

Proof. (a) Suppose that $f'(x) \geq 0$ for all $x \in I$.

If x_1, x_2 in I satisfy $x_1 < x_2$, then we apply the Mean Value theorem to f on the closed interval $J := [x_1, x_2]$ to obtain a point c in (x_1, x_2) such that

$$f(x_2) - f(x_1) = f'(c)(x_2 - x_1).$$

Since $f'(c) \geq 0$ and $x_2 - x_1 > 0$, it follows that $f(x_2) - f(x_1) \geq 0$.

Hence,

$$f(x_1) \leq f(x_2)$$

and, since $x_1 < x_2$ are arbitrary points in I, we conclude that f is increasing on I.

Conversely, suppose that f is differentiable and increasing on I.

Now, for any point $x \neq c$ in I, we have

$$(f(x) - f(c))/(x - c) \geq 0.$$

Therefore, we conclude that

$$f'(c) = \lim_{x \to c} \frac{f(x) - f(c)}{x - c} \geq 0.$$

Since, c is arbitrary, therefore, this proves that $f'(x) \geq 0$ for all $x \in I$.

(b) The proof of part (b) is similar.

Remark 1.6. *A function f is said to be strictly increasing on an interval I if for any points x_1, x_2 in I such that $x_1 < x_2$, we have $f(x_1) < f(x_2)$.*

A function f is said to be strictly decreasing on an interval I if for any points x_1, x_2 in I such that $x_1 < x_2$, we have $f(x_1) > f(x_2)$.

Theorem 1.13. First Derivative Test for Extrema

Let f be continuous on the interval $I := [a, b]$ and let c be an interior point of I. Assume that f is differentiable on (a, c) and (c, b). Then

(a) If there is a neighborhood $(c - \delta, c + \delta) \subseteq I$ such that $f'(x) \geq 0$ for $c - \delta < x < c$ and $f'(x) \leq 0$ for $c < x < c + \delta$ then f has a relative maximum at c.

(b) If there is a neighborhood $(c - \delta, c + \delta) \subseteq I$ such that $f'(c) \leq 0$ for $c - \delta < x < c$ and $f'(x) \geq 0$ for $c < x < c + \delta$, then f has a relative minimum at c.

Proof. (a) If $x \in (c - \delta, c)$, then it follows from the Mean Value theorem that there exists a point $c_x \in (x, c)$ such that

$$f(c) - f(x) = (c - x)f'(c_x).$$

Since, $f'(c_x) \geq 0$ therefore, we conclude that

Similarly, it follows that

$$f(x) \leq f(c) \quad \text{for} \quad x \in (c, c + \delta).$$

Therefor,e $f(x) \leq f(c)$ for all $x \in (c - \delta, c + \delta)$, so that f has a relative maximum at c.

(b) The proof is similar.

The Intermediate Value Property of Derivatives

Lemma 1.1. *Let $I \subseteq \mathbb{R}$ be an interval, let $f : I \to \mathbb{R}$, let $c \in I$, and assume that f has a derivative at c. Then*

(a) If $f'(c) > 0$, then there is a number $\delta > 0$ such that $f(x) > f(c)$ for $x \in I$ such that $c < x < c + \delta$.

(b) If $f'(c) < 0$, then there is a number $\delta > 0$ such that $f(x) > f(c)$ for $x \in I$ such that $c - \delta < x < c$.

Proof. (a) Since

$$\lim_{x \to c} \frac{f(x) - f(c)}{x - c} = f'(c) > 0.$$

Therefore, there is a number $\delta > 0$ such that if $x \in I$ and $0 < |x - c| < \delta$, then

$$\frac{f(x) - f(c)}{x - c} > 0.$$

If $x \in I$ also satisfies $x > c$, then we have

$$f(x) - f(c) = (x - c) \cdot \frac{f(x) - f(c)}{x - c} > 0.$$

Hence, if $x \in I$ and $c < x < c + \delta$, then $f(x) > f(c)$.

The proof of (b) is similar.

Illustrative Examples

1. For each of the following functions on $\mathbb{R}$ to $\mathbb{R}$, find points of relative extrema, the intervals on which the function is increasing, and those on which it is decreasing

   (i) $f(x) = x^2 - 3x + 5,$ (ii) $g(x) = 3x - 4x^2$.

   Solution: (i) Here $f(x) = x^2 - 3x + 5$.

   The derivative of the given function f is $f'(x) = 2x - 3$. Let c be an extreme point of f.

   Therefore, by Interior Extremum theorem (3.70) it follows that $f'(c) = 0$;

   $$\therefore f'(c) = 0 \Rightarrow 2c - 3 = 0 \Rightarrow c = \frac{3}{2}$$

   Therefore, $c = \frac{3}{2}$ is a point of relative extrema and the intervals are $\left(-\infty, \frac{3}{2}\right], \left[\frac{3}{2}, \infty\right)$.

   For $x \geq c$, we have

   $$2x - 3 \geq 0 \Rightarrow f'(x) \geq 0.$$

   Therefore, by theorem (1.12), we conclude that f is increasing on interval $\left[\frac{3}{2}, \infty\right)$.

   For $x \leq c$, we have

Therefore, by theorem (1.12), we conclude that f is decreasing on interval $\left(-\infty, \frac{3}{2}\right]$.

(ii) Here $g(x) = 3x - 4x^2$.

The derivative of the given function is $g'(x) = 3 - 8x$. Let c be an extreme point of g.

Therefore, by Interior Extremum theorem (3.70) it follows that $g'(c) = 0$;

$$\therefore g'(c) = 0 \Rightarrow 3 - 8c = 0 \Rightarrow c = \frac{3}{8}.$$

Therefore, $c = \frac{3}{8}$ is a point of relative extrema.

For $x \geq c$, we have

$$3 - 8x \leq 0 \Rightarrow g'(x) \leq 0.$$

Therefore, by theorem (1.12), we conclude that g is decreasing on interval $\left[\frac{3}{8}, \infty\right)$.

For $x \leq c$, we have

$$3 - 8x \geq 0 \Rightarrow g'(x) \geq 0.$$

Therefore, by theorem (1.12), we conclude that g is increasing on interval $\left(-\infty, \frac{3}{8}\right]$.

2. Find the points of relative extrema, the intervals on which the following functions are increasing, and those on which they are decreasing:

(i) $f(x) = x + \dfrac{1}{x}$ for $x \neq 0$ (ii) $g(x) = \dfrac{x}{(x^2 + 1)}$ for $x \in \mathbb{R}$.

Solution: (i) Here $f(x) = x + \frac{1}{x}, x \neq 0$. Therefore, f is differentiable on $\mathbb{R} - \{0\}$

Now,

$$f'(x) = 1 - \frac{1}{x^2}.$$

$$\therefore f'(x) = 0 \Rightarrow x = \pm 1.$$

Then 1 and -1 are points of relative extrema.

If $0 < |x| \leq 1$ then

$$\frac{1}{|x|} \geq 1 \Rightarrow \frac{1}{x^2} - 1 \geq 0 \Rightarrow f'(x) \leq 0.$$

Therefore, on $[-1, 0)$, $f'(x) \leq 0$ and on $(0, 1]$, $f'(x) \leq 0$.

Hence, f is decreasing on $[-1, 0)$ and on $(0, 1]$.

For $x \geq 1, 1 - \dfrac{1}{x^2} \geq 0 \Rightarrow f'(x) \geq 0.$

Therefore, f is increasing on $[1, \infty)$.

For $x \geq 1, 1 - \dfrac{1}{x^2} \geq 0 \Rightarrow f'(x) \geq 0$

Therefore, f is increasing on $(-\infty, -1]$.

(ii) Here $g(x) = \dfrac{x}{x^2 + 1}, x \in \mathbb{R}$.

Then g is differentiable on $\mathbb{R}$, and

$$g'(x) = \frac{1 + x^2 - x \cdot 2x}{(1 + x^2)^2} = \frac{1 - x^2}{(1 + x^2)^2}.$$

Then $g'(x) = \Rightarrow 1 - x^2 = 0 \Rightarrow x = \pm 1$.

Therefore, 1 and -1 are points of relative extrema.

For $-1 \leq x \leq 1 \Rightarrow 1 - x^2 \geq 0$.

Therefore, $g'(x) \geq 0$ on $[-1, 1]$.

Hence, g is increasing on $[-1, 1]$.

If $x \geq 1$ then $1 - x^2 \leq 0$ so $g'(x) \leq 0$.

Therefore g is decreasing on $[1, \infty)$.

Also, for $x \leq -1, 1 - x^2 \leq 0$ then $g'(x) \leq 0$.

Hence, g is decreasing on $(-\infty, -1]$.

3. Find the points of relative extrema of the following functions on the specified domain:

   (a) $f(x) = |x^2 - 1|$ for $-4 \leq x \leq 4$ (b) $g(x) = 1 - (x - 1)^{2/3}$ for $0 \leq x \leq 2$.

   Solution: (a) Here $f(x) = |x^2 - 1|$ $x \in [-4, 4]$

   Hence,

   $$f(x) = \begin{cases} x^2 - 1, & \text{for} \quad x \in [-4, -1] \cup [1, 4], \\ -1 - x^2, & \text{for} \quad x \in [-1, 1]. \end{cases}$$

   Therefore, f is differentiable on interior of each of the above intervals $[-4, -1], [1, 4], [-1, 1]$.

   We have

   $$f'(x) = \begin{cases} 2x, & \text{for} \quad x \in (-4, -1) \cup (1, 4), \\ -2x, & \text{for} \quad x \in (-1, 1). \end{cases}$$

   Therefore,

   $$f'(x) > 0 \text{ for } x \in (1, 4)$$

   $$f'(x) < 0 \text{ for } x \in (-4, -1)$$

   Also,

   $$f'(x) > 0 \text{ for } x \in (-1, 0)$$

   $$f'(x) < 0 \text{ for } x \in (0, 1).$$

   Hence, f is increasing on $(1, 4)$, decreasing on $(-4, -1)$, increasing on $(-1, 0)$ and decreasing on $(0, 1)$. Thus the points of relative extrema are $x = -4, -1, 0, 1, 4$.

   (b) Here $g(x) = 1 - (x - 1)^{\frac{2}{3}}$ for $0 \leq x \leq 2$.

   $$g'(x) = -\frac{2}{3}(x - 1)^{-\frac{1}{3}} = -\frac{2}{3(x - 1)^{\frac{1}{3}}}.$$

   Therefore,

   $$g'(x) > 0 \text{ for } x \in (0, 1)$$

Therefore, g is increasing on $(0,1)$ and decreasing on $(1,2)$, g is continuous as well. Therefore, the points of extrema are $x = 0, 1, 2$.

4. Use the Mean Value Theorem to prove that $|\sin x - \sin y| \leq |x - y|$ for all x, y in $\mathbb{R}$.

Solution: Let $f(x) = sinx$.

Now, sine function is continuous and differentiable on $\mathbb{R}$.

By Mean Value theorem we know that for a function which is continuous on a closed interval $[a, b]$ and that has a derivative in the open interval (a, b) there exists at least one point c in (a, b) such that

$$f(b) - f(a) = f'(c)(b - a).$$

Therefore, f satisfies the above mentioned criterion, so by applying the theorem, we get

$$
\begin{aligned}
|\sin x - \sin y| &= |\cos c(x - y)| \\
\frac{|\sin x - \sin y|}{|x - y|} &= |\cos c|
\end{aligned}
$$

Now, we know that it is always true that $|\cos x| \leq 1 \quad \forall\, x$, which means that

$$
\begin{aligned}
\frac{|\sin x - \sin y|}{|x - y|} &\leq 1 \\
\Rightarrow |\sin x - \sin y| &\leq |x - y|.
\end{aligned}
$$

5. Use the Mean Value Theorem to prove that $\dfrac{(x-1)}{x} < \ln x < x - 1$, for $x > 1$.

Solution: Let $f(x) = \ln x - \dfrac{x-1}{x} = \ln x - 1 + \dfrac{1}{x}, x \geq 1$

Then f is differentiable on $[1, \infty)$ and

$$
\begin{aligned}
f'(x) &= \frac{1}{x} - \frac{1}{x^2} \\
&= \frac{x-1}{x^2} > 0 \text{ for } x > 1
\end{aligned}
$$

Therefore $f'(x) > 0$ for $x > 1$.

Hence, f is strictly increasing on $[1, \infty)$.

Therefore, $f(x) > f(1)$, $\forall\, x > 1$.

$$\Rightarrow \ln x - \frac{x-1}{x} > 0, \ \forall\, x > 1 \ \ as \ \ f(1) = 0.$$

$$\Rightarrow \frac{x-1}{x} < \ln x, \quad \text{for } x > 1 \tag{1.2}$$

Let $h(x) = x - 1 - \ln x, \quad x \geq 1$.

Then h is differentiable on $[1, \infty)$ and

$$h'(\ ) = 1 \qquad 1 \qquad x - 1 \qquad 0, \ \ for \ \ x > 1$$

Therefore h is strictly increasing on $[1, \infty)$ by theorem (1.12).

Then $h(x) > h(1)$ for all $x > 1$

But $h(1) = 0$, therefore, $h(x) > 0$, $\forall\, x > 1$.

$$\Rightarrow \ln x < x - 1, \quad \forall\, x > 1. \tag{1.3}$$

From equations (1.2) and (1.3), we get

$$\frac{x - 1}{x} < \ln x < x - 1, \quad \forall\, x > 1.$$

6. Let $f : [a, b] \to \mathbb{R}$ be continuous on $[a, b]$ and differentiable in (a, b). Show that if $\lim\limits_{x \to a} f'(x) = A$, then $f'(a)$ exists and equals A.

   Solution: By definition, the derivative of f at a is given by

   $$f'(a) = \lim_{x \to a} \frac{f(x) - f(a)}{x - a}.$$

   provided this limit exists.

   By Mean value Theorem we know that for any $x \in (a, b)$, where $a < c < x$

   $$f(x) - f(a) = f'(c)(x - a).$$

   Therefore,
   $$\frac{f(x) - f(a)}{x - a} = f'(c).$$

   If $x \to a$ it follows that $c \to a$ adn $\lim\limits_{c \to a} f'(x)$ exists and is equal to A.

   Therefore,
   $$\lim_{x \to a} \frac{f(x) - f(a)}{x - a} = \lim_{c \to a} f'(x) = A.$$

   Hence, $f'(a)$ exists and equal to A.

7. Let $g : \mathbb{R} \to \mathbb{R}$ be defined by

   $$g(x) = \begin{cases} x + 2x^4 \sin\left(\frac{1}{x}\right), & \text{for} \quad x \neq 0 \\ 0, & \text{for} \quad x = 0 \end{cases}$$

   Show that $g'(0) = 1$ but in every neighborhood of 0 the derivative $g'(x)$ takes on both positive and negative values. Thus g is not monotonic in any neighborhood of 0.

   Solution: Here

   $$g(x) = \begin{cases} x + 2x^4 \sin\left(\frac{1}{x}\right), & \text{for} \quad x \neq 0 \\ 0, & \text{for} \quad x = 0. \end{cases}$$

To show that $g'(0) = 1$.

$$\begin{aligned}
g'(0) &= \lim_{x \to 0} \frac{f(x) - f(0)}{x - 0} \\
&= \lim_{x \to 0} \frac{x + 2x^2 \sin \frac{1}{x}}{x} \\
&= \lim_{x \to 0} \frac{x(1 + 2x \sin \frac{1}{x})}{x} \\
&= \lim_{x \to 0} (1 + 2x \sin \frac{1}{x}) \\
&= \lim_{x \to 0} 1 + \lim_{x \to 0} 2x \sin \frac{1}{x} \\
&= 1 + 0 = 1
\end{aligned}$$

We have prove that $g'(0) = 1$.

For $x \neq 0$ $g'(x) = 1 + 4x \sin \frac{1}{x} - 2 \cos \frac{1}{x}$.

Let $x_n = \frac{1}{2n\pi}, x_n \to 0$, we have

$$g'(x_n) = 1 + 0 - 2 = -1 < 0.$$

Let $y_n = \frac{1}{(4n+1)\frac{\pi}{2}}, y_n \to 0$, we have

$$g'(y_n) = 1 + 4\frac{1}{(4n+1)\frac{\pi}{2}} - 0 > 0.$$

Therefore, g' takes both positive and negative values in neighborhood of 0. In other words g is not monotonic in any neighborhood of 0.

8. Let I be an interval. Prove that if f is differentiable on I and if the derivative f' is bounded on I, then f satisfies a Lipschitz condition on I.

   Solution: Now, $f : I \to \mathbb{R}$ is differentiable f' is bounded on I. To prove f' satisfies Lipchitz condition on I.

   Let $M > 0$ be such that

   $$|f'(c)| \leq M \ \ \forall \, c \in I$$

   because f' is bounded on I.

   For $x < y, x, y \in I$, by mean value theorem, we get $c \in (x, y) \subseteq I$, such that

   $$f(y) - f(x) = f'(c)(y - x).$$

   $\Rightarrow |f(x) - f(y)| = |f'(c)||y - x| \leq M|x - y|, \ \ c \in I$

   $\Rightarrow |f(x) - f(y)| \leq M|x - y| \ \ \forall \, x, y, \in I.$

   (Note that $M > 0$ is independent of x, y). Hence, f satisfies Lipchitz condition on I.

Exercise

1. Determine whether the following functions are differentiable at $x = 0$.

   (i) $f(x) = \begin{cases} x^2, & if \ x \geq 0 \\ -x^2, & if \ x < 0 \end{cases}$ (ii) $f(x) = \begin{cases} x^3, & if \ x \geq 0 \\ 0, & if \ x < 0 \end{cases}$

2. Determine whether the following functions are differentiable at $x = 1$.

   (i) $f(x) = \begin{cases} x^2, & if \ x \leq 1 \\ x, & if \ x > 0 \end{cases}$ (ii) $f(x) = \begin{cases} x^3, & if \ x \leq 0 \\ 4x - 3, & if \ x > 0 \end{cases}$

3. Use the definition to find the derivative of each of the following functions.

   (i) $f(x) = 2x^3 + 3x + 1$ for $x \in \mathbb{R}$,   (ii) $g(x) = \frac{1}{\sqrt{2-x}}$ for $x \in \mathbb{R}, x \neq 2$.

4. Recalling the definition of the derivative evaluate $\lim\limits_{h \to 0} \frac{(3+h)^5 - 3^5}{h}$.

5. Calculate the derivative of the following functions.

   (i) $(x^4 - 3x^2 + 5x - 2)^3$, (ii) $\sqrt{7x^3 - 2x^2 + 5}$, (iii) $\frac{1}{(5x^2+4)^3}$.

6. Verify Rolle's theorem for following functions, if so, find the value of c in the open interval (a, b) such that $f'(c) = 0$.

   (i) $f(x) = \frac{x^2 - 4x + 3}{x - 2}$, on $[1, 3]$,   (ii) $f(x) = |x - 1|$, on $[0, 2]$

   (iii) $f(x) = (x - 1)(x - 2)(x - 3)$, on $[1, 3]$,   (iv) $f(x) = ln(2 - x^2)$, on $[-1, 1]$

   (v) $f(x) = sech x = \frac{2}{e^x + e^{-x}}$, on $[-1, 1]$,   (vi) $f(x) = sin(x/2)$, on $[0, 2\pi]$.

7. Verify Mean Value theorem for following functions, if so, find the value of c in the open interval (a, b).

   (i) $f(x) = \frac{x}{x+1}$, on $[1, 3]$,   (ii) $f(x) = \sqrt{x}$, on $[1, 9]$

   (iii) $f(x) = cos x + tan x$, on $[-\pi, \pi]$,   (iv) $f(x) = x - 2sin x$, on $[-\pi, \pi]$

   (v) $f(x) = x^4 - 2x^3 + x^2$, on $[0, 6]$,   (vi) $f(x) = x^{\frac{2}{3}}$, on $[0, 1]$.

8. Let $f(x)$ be differentiable function for all $x \in \mathbb{R}$. suppose that $f(0) = -3$ and $f'(x) \leq 5$ for all $x \in \mathbb{R}$. Use Mean Value theorem to show that $f(2) \leq 7$.

9. Verify Cauchy's Mean Value theorem for following functions, if so, find the value of c in the open interval (a, b).

   (i) $f(x) = x^4, g(x) = x^2$, on $[a, b], [1, 2]$ (ii) $f(x) = x^3, g(x) = tan^{-1}x$, on $[0, 1]$,

   (iii) $f(x) = cos x, g(x) = sin x$, on $[a, b]$.

10. Using Cauchy's Mean Value theorem show that $\lim\limits_{x \to 0} \frac{sin x}{x} = 1$.

11. For each of the following functions on $\mathbb{R}$ to $\mathbb{R}$, find points of relative extrema, the intervals on which the function is increasing, and those on which it is decreasing.

   (i) $f(x) = x^3 - 3x - 4$,   (ii) $g(x) = x^4 + 2x^2 - 4$.

12. Find the points of relative extrema, the intervals on which the following functions are increasing, and those on which they are decreasing

    (i) $f(x) = \sqrt{x} - 2\sqrt{x+2}$ for $x > 0$,   (ii) $g(x) = 2x + 1/x^2$ for $x \neq 0$.

13. Find the points of relative extrema of the following functions on the specified domain:

    (i) $h(x) = x|x^2 - 12|$ for $-2 \leq x \leq 3$,   (ii) $k(x) = x(x-8)^{1/3}$ for $0 \leq x \leq 9$.

Solutions

1. (i) ,(ii) differentiable at $x = 0$.

2. (i) ,(ii) not differentiable at $x = 1$.

3. (i) $6x^2 + 3$,(ii) $\frac{1}{2(2-x)^{3/2}}$.

6. (i), (ii) Not satisfied, (iv), (v) $c = 0$ (vi) $c = \pi$.

7. (i) $c = -1 + 2\sqrt{2}$, (ii) $c = 4$ot, (iv) $c = \frac{\pi}{2}$, (vi) $c = \frac{8}{27}$.

9. (i) $c = \pm\sqrt{\frac{a^2+b^2}{2}}$, (ii) $c = \sqrt{\frac{\pi}{12}} - \pi$, (iii) $c = \frac{a+b}{2}$.

11. (i) decreasing on $[-1, 1]$, increasing on $(-\infty, -1) \cup (1, \infty)$, (ii) increasing on $[0, \infty)$, decreasing on $(-\infty, 0]$.

12. (i) increasing on $(0, 2/3]$, decreasing on $[2/3, \infty)$, (ii) increasing on $(-\infty, -1], [1, \infty)$, decreasing on $[-1, 0), (0, 1]$.

Chapter 2

Unit 2: L' Hospital Rule and Successive Differentiation

2.1 Introduction

In this section, we will discuss limit theorems that involve cases that cannot be determined by previous limit theorems. For example, if $f(x)$ and $g(x)$ both approach 0 as x approaches a, then the quotient $\frac{f(x)}{g(x)}$ may or may not have a limit at a and it is said to have the indeterminate form $0/0$. The limit theorem for this case is due to Johann Bernoulli and first appeared in the 1696 book published by L'Hospital. **Johann Bernoulli** (16671748) was born in Basel, Switzerland. In the year 1692, Johann met the Marquis Guillame Francois de L'Hospital and agreed to a financial arrangement under which he would teach the new calculus to L'Hospital, giving L'Hospital the right to use Bernoullis lessons as he pleased. The initial theorem was refined and extended, and the various results are collectively referred to as LHospitals (or L'Hôpital's) Rules. In this section, we establish some basic results that is based simply on the definition of the derivative.

Theorem 2.1. *Let f and g be defined on $[a, b]$, let $f(a) = g(a) = 0$, and let $g(x) \neq 0$ for $a < x < b$. If f and g are differentiable at a and if $g'(a) \neq 0$, then the limit of $\frac{f}{g}$ at a exists and is equal to $\frac{f'(a)}{g'(a)}$. Thus*

$$\lim_{x \to a+} \frac{f(x)}{g(x)} = \frac{f'(a)}{g'(a)}.$$

Proof. Since $f(a) = g(a) = 0$, we can write the quotient $\frac{f(x)}{g(x)}$ for $a < x < b$ as follows

$$\frac{f(x)}{g(x)} = \frac{f(x) - f(a)}{g(x) - g(a)} = \frac{\dfrac{f(x) - f(a)}{x - a}}{\dfrac{g(x) - g(a)}{x - a}}.$$

Taking limit as $x \to a^+$, we obtain

$$\lim_{x \to a+} \frac{f(x)}{g(x)} = \frac{\displaystyle\lim_{x \to a+} \frac{f(x) - f(a)}{x - a}}{\displaystyle\lim_{x \to a+} \frac{g(x) - g(a)}{x - a}} = \frac{f'(a)}{g'(a)}$$

$$\lim_{x \to a+} \frac{f(x)}{g(x)} = \frac{f'(a)}{g'(a)}.$$

2.2   L'Hospital's Rules

The result asserts that the limiting behavior of $\frac{f(x)}{g(x)}$ as $x \to a+$ is the same as the limiting behavior of $\frac{f'(x)}{g(x)}$ as $x \to a+$, including the case where this limit is infinite. An important hypothesis here is that both f and g approach 0 as $x \to a +$.

Theorem 2.2. L'Hospital's Rule- I

Let $-\infty \leq a < b \leq \infty$ and let f, g be differentiable on (a, b) such that $g'(x) \neq 0$ for all $x \in (a, b)$. Suppose that $\lim\limits_{x \to a+} f(x) = 0 = \lim\limits_{x \to a+} g(x)$.

(i) If $\lim\limits_{x \to a+} \dfrac{f'(x)}{g'(x)} = L \in \mathbb{R}$, then $\lim\limits_{x \to a+} \dfrac{f(x)}{g(x)} - L$.

(ii) If $\lim\limits_{x \to a+} \dfrac{f'(x)}{g'(x)} = L \in \{-\infty, \infty\}$, then $\lim\limits_{x \to a+} \dfrac{f(x)}{g(x)} = L$.

(iii) If $\lim\limits_{x \to a+} \dfrac{f''(x)}{g''(x)} = L$ then $\lim\limits_{x \to a+} \dfrac{f(x)}{g(x)} = L$.

Example 2.1. *(i) We consider $\lim\limits_{x \to 0} \dfrac{\sin x}{\sqrt{x}}$.*
Let $f(x) = \sin x$ and $g(x) = \sqrt{x}$, we have

$$\lim_{x \to 0+} \frac{f(x)}{g(x)} = \lim_{x \to 0+} \frac{\sin x}{\sqrt{x}} \quad (form \ \ \frac{0}{0}$$

Therefore, by applying L'Hospital's Rule, we get

$$\lim_{x \to 0+} \frac{f(x)}{g(x)} = \lim_{x \to 0+} \left[\frac{\cos x}{1/(2\sqrt{x})} \right]$$
$$= \lim_{x \to 0+} 2\sqrt{x} \cos x$$
$$= 0.$$

Observe that the denominator is not differentiable at $x = 0$ so that theorem (2.1) cannot be applied. However

$$f(x) = \sin x \quad and \quad g(x) = \sqrt{x}$$

are differentiable on $(0, 1)$ and both approach 0 as $x \to 0+$. Moreover, $g'(x) \neq 0$ on $(0, 1)$, so that (2.1) is applicable.

(ii) We consider $\lim\limits_{x\to 0} \dfrac{1-\cos x}{x^2}$.

Let $f(x) = 1 - \cos x$ and $g(x) = x^2$, we have

$$\lim_{x\to 0+} \frac{f(x)}{g(x)} = \lim_{x\to 0+} \frac{1-\cos x}{x^2} \qquad (\text{form } \frac{0}{0}$$

Therefore, by applying L'Hospital's Rule, we get

$$\lim_{x\to 0+} \frac{f(x)}{g(x)} = \lim_{x\to 0+} \left[\frac{1-\cos x}{x^2}\right]$$
$$= \lim_{x\to 0+} \frac{\sin x}{2x}$$
$$= \frac{1}{2} \lim_{x\to 0} \frac{\sin x}{x}$$
$$= \frac{1}{2}.$$

(iii) We consider $\lim\limits_{x\to 0} \dfrac{e^x - 1 - x}{x^2}$.

Let $f(x) = e^x - 1 - x$ and $g(x) = x^2$, we have

$$\lim_{x\to 0} \frac{f(x)}{g(x)} = \lim_{x\to 0} \frac{e^x - 1 - x}{x^2} \quad (\text{form}\frac{0}{0}$$

Therefore, by applying L'Hospital's Rule, we get

$$\lim_{x\to 0} \frac{f(x)}{g(x)} = \lim_{x\to 0+} \left[\frac{e^x - 1 - x}{x^2}\right]$$
$$= \lim_{x\to 0} \frac{e^x - 1}{2x} \qquad (\text{form } \frac{0}{0}$$
$$= \lim_{x\to 0} \frac{e^x}{2}$$
$$= \frac{1}{2}.$$

(iv) We consider $\lim\limits_{x\to 0} \dfrac{\ln x}{x - 1}$.

Let $f(x) = \ln x$ and $g(x) = x - 1$, we have

$$\lim_{x\to 1} \frac{f(x)}{g(x)} = \lim_{x\to 1} \frac{\ln x}{x - 1} \quad (\text{form}\frac{0}{0}$$

Therefore, by applying L'Hospital's Rule, we get

$$\lim_{x\to 1} \frac{f(x)}{g(x)} = \lim_{x\to 1} \left[\frac{\ln x}{x - 1}\right]$$
$$= \lim_{x\to 1} \frac{\left(\frac{1}{x}\right)}{1}$$
$$= 1.$$

Theorem 2.3. L'Hospital's Rule- II

Let $-\infty \le a < b \le \infty$ and let f, g be differentiable on (a, b) such that $g'(x) \ne 0$ for all $x \in (a, b)$.

(i) If $\lim\limits_{x\to a+}\dfrac{f'(x)}{g'(x)} = L \in \mathbb{R},$ *then* $\lim\limits_{x\to a+}\dfrac{f(x)}{g(x)} - L.$

(ii) If $\lim\limits_{x\to a+}\dfrac{f'(x)}{g'(x)} = L \in \{-\infty, \infty\},$ *then* $\lim\limits_{x\to a+}\dfrac{f(x)}{g(x)} = L.$

(iii) If $\lim\limits_{x\to a+}\dfrac{f''(x)}{g''(x)} = L$ *then* $\lim\limits_{x\to a+}\dfrac{f(x)}{g(x)} = L.$

Example 2.2. *(a) We consider* $\lim\limits_{x\to\infty}\dfrac{\ln x}{x}.$

Let $f(x) = \ln x$ *and* $g(x) = x$ *on the interval* $(0, \infty)$. *If we apply the left-hand version of theorem (2.3), we obtain*

$$\lim_{x\to\infty}\frac{f(x)}{g(x)} = \lim_{x\to\infty}\frac{\ln x}{x} \quad (form\frac{\infty}{\infty})$$

Therefore, by applying L'Hospital's Rule, we get

$$\lim_{x\to\infty}\frac{\ln x}{x} = \lim_{x\to\infty}\frac{1/x}{1}$$
$$= 0.$$

(b) We consider $\lim\limits_{x\to\infty} e^{-x}x^2.$

Here we take $f(x) = x^2$ and $g(x) = e^x$ on $\mathbb{R}$.

$$\lim_{x\to\infty}\frac{f(x)}{g(x)} = \lim_{x\to\infty}\frac{x^2}{e^x} \quad (\text{form}\frac{\infty}{\infty})$$

Therefore, by applying L'Hospital's Rule, we get

$$\lim_{x\to\infty}\frac{x^2}{e^x} = \lim_{x\to\infty}\frac{2x}{e^x} \quad (\text{form}\frac{\infty}{\infty})$$
$$= \lim_{x\to\infty}\frac{2}{e^x}$$
$$= 0.$$

(c) We consider $\lim\limits_{x\to 0+}\dfrac{\ln\sin x}{\ln x}$

Here we take $f(x) := \ln\sin x$ and $g(x) := \ln x$ on $(0, \pi)$.

$$\lim_{x\to 0+}\frac{f(x)}{g(x)} = \lim_{x\to 0+}\frac{\ln\sin x}{\ln x} \quad (\text{form}\frac{\infty}{\infty})$$

Therefore, by applying L'Hospital's Rule, we get

$$\lim_{x\to 0+}\frac{\ln\sin x}{\ln x} = \lim_{x\to 0+}\frac{x}{\sin x}\cos x$$
$$= 1.$$

Other Indeterminate Forms

Indeterminate forms such as $\infty - \infty, 0 \cdot \infty, 1^\infty, 0^0, \infty^\circ$ can be reduced to the previously considered cases by algebraic manipulations and the use of the logarithmic and exponential functions. We

Example 2.3. *Evaluate* $\lim\limits_{x\to 0+}\left(\dfrac{1}{x}-\dfrac{1}{\sin x}\right)$, $I=(0,\pi/2)$.

Solution: Consider

$$\lim_{x\to 0+}\left(\frac{1}{x}-\frac{1}{\sin x}\right)$$

which has the indeterminate form $\infty-\infty$.

We have

$$
\begin{aligned}
\lim_{x\to 0+}\left(\frac{1}{x}-\frac{1}{\sin x}\right) &= \lim_{x\to 0+}\frac{\sin x - x}{x\sin x} && \left(\text{form}\frac{0}{0},\ \text{Apply L'Hospitals rule}\right.\\[2mm]
&= \lim_{x\to 0+}\frac{\cos x - 1}{\sin x + x\cos x}\\[2mm]
&= \lim_{x\to 0+}\frac{-\sin x}{2\cos x - x\sin x}\\[2mm]
&= \frac{0}{2}\\[2mm]
&= 0.
\end{aligned}
$$

Example 2.4. *Evaluate* $\lim\limits_{x\to 0+} x\ln x$, $I=(0,\infty)$.

Solution: Let $I:=(0,\infty)$ and consider

$$\lim_{x\to 0+} x\ln x$$

which has the indeterminate form $0\cdot(-\infty)$.

We have

$$
\begin{aligned}
\lim_{x\to 0+} x\ln x &= \lim_{x\to 0+}\frac{\ln x}{1/x} && \left(\text{form}\frac{\infty}{\infty},\text{Apply L'Hospitals rule}\right.\\[2mm]
&= \lim_{x\to 0+}\frac{1/x}{1/x^2}\\[2mm]
&= \lim_{x\to 0+}(-x)\\[2mm]
&= 0.
\end{aligned}
$$

Illustrative Examples

1. Suppose that f and g are continuous on $[a,b]$, differentiable on (a,b), that $c\in[a,b]$ and that $g(x)\neq 0$ for $x\in[a,b], x\neq c$. Let $A:=\lim\limits_{x\to c} f$ and $B:=\lim\limits_{x\to c} g$. If $B=0$, and if $\lim\limits_{x\to c} f(x)/g(x)$ exists in R, show that we must have $A=0$.

   Solution: If $x\neq c$ then we can write $f(x)$ as

$$f(x)=\frac{f(x)}{g(x)}\cdot g(x).$$

Taking limit, we obtaon

$$
\begin{aligned}
\lim_{x \to c} f(x) &= \lim_{x \to c} \left(\frac{f(x)}{g(x)} \cdot g(x) \right) \\
&= \lim_{x \to c} \frac{f(x)}{g(x)} \cdot \lim_{x \to c} g(x) \\
&= \lim_{x \to c} \frac{f(x)}{g(x)} \cdot B \\
&= 0
\end{aligned}
$$

This prove $A = \lim_{x \to c} f(x) = 0$.

2. Let

$$
f(x) = \begin{cases} x^2 \sin\left(\frac{1}{x}\right), & for \ 0 < x \le 1 \\ 0, & x = 0. \end{cases}
$$

and let $g(x) := x^2$ for $x \in [0, 1]$. Then both f and g are differentiable on $[0, 1]$ and $g(x) > 0$ for $x \ne 0$. Show that $\lim_{x \to 0} f(x) = 0 = \lim_{x \to 0} g(x)$ and that $\lim_{x \to 0} \frac{f(x)}{g(x)}$ does not exist.

Solution: Now, $|f(x)| = |x^2 \sin \frac{1}{x}| \le |x^2|$.

We have

$$
-|x^2| \le |f(x)| \le |x^2|.
$$

We know that $\lim_{x \to 0} x^2 = 0$ and $\lim_{x \to 0} -x^2 = 0$, therefore, by squeeze theorem, we get

$$
\lim_{x \to 0} f(x) = 0
$$

For $g(x)$ we have:

$$
\lim_{x \to 0} g(x) = \lim_{x \to 0} x^2 = 0
$$

This proves $\lim_{x \to 0} f(x) = 0 = \lim_{x \to 0} g(x)$.

For $x \ne 0$

$$
\frac{f(x)}{g(x)} = \frac{x^2 \sin \frac{1}{x}}{x^2} = \sin \frac{1}{x}.
$$

Let $(x_n), (y_n)$ be sequences defined as $x_n = \frac{1}{2n\pi}$, $y_n = \frac{1}{(2n+1)\frac{\pi}{2}}$ and both are converge to zero as $n \to \infty$.

Now, $\sin \frac{1}{x_n} = 0$, $\sin \frac{1}{y_n} = \pm 1$.

Therefore, these are different values hence, limit does not exists.

Therefore, $\lim_{x \to 0} \frac{f(x)}{g(x)}$ does not exist.

3. Let $f(x) := x^2 \sin\left(\frac{1}{x}\right)$ for $x \ne 0$, let $f(0) := 0$, and let $g(x) := \sin x$ for $x \in \mathbb{R}$. Show that $\lim_{x \to 0} \frac{f(x)}{g(x)} = 0$ but that $\lim_{x \to 0} \frac{f'(x)}{g'(x)}$ does not exist.

Solution: To calculate the limit $\lim\limits_{x\to 0}\dfrac{f(x)}{g(x)}$.

$$
\begin{aligned}
\lim_{x\to 0}\frac{f(x)}{g(x)} &= \lim_{x\to 0}\frac{x^2 \sin\left(\frac{1}{x}\right)}{\sin(x)} \\[2mm]
&= \lim_{x\to 0}\frac{x}{\sin(x)} \cdot \lim_{x\to 0} x\sin\left(\frac{1}{x}\right) \\[2mm]
&= 1\cdot 0 \\[2mm]
&= 0
\end{aligned}
$$

We have shown that $\lim\limits_{x\to 0}\dfrac{f(x)}{g(x)} = 0$.

Now, lets see if we can calculate the limit of $\lim\limits_{x\to 0}\dfrac{f'(x)}{g'(x)}$

$$
\begin{aligned}
\lim_{x\to 0}\frac{f'(x)}{g'(x)} &= \lim_{x\to 0}\frac{2x\sin\left(\frac{1}{x}\right)\left(-\cos\frac{1}{x}\right)}{\cos(x)} \\[2mm]
&= \lim_{x\to 0}\frac{2x\sin\left(\frac{1}{x}\right)}{\cos(x)} \cdot \left(-\lim_{x\to 0}\cos\frac{1}{x}\right) \quad \text{does not exists.}
\end{aligned}
$$

Therefore ,the limit $\lim\limits_{x\to 0}\dfrac{f'(x)}{g'(x)}$ does not exists.

4. Evaluate the following limits, where the domain of the quotient is as indicated.

(i) $\lim\limits_{x\to 0+}\dfrac{\ln(x+1)}{\sin x}$ $(0,\pi/2)$,   (ii) $\lim\limits_{x\to 0+}\dfrac{\tan x - x}{x^3}$.

Solution: (i) Consider $\lim\limits_{x\to 0^+}\dfrac{\ln(1+x)}{\sin x}$, $\left(0,\dfrac{\pi}{2}\right)$ which has the indeterminate form $\frac{0}{0}$.

We have

$$
\begin{aligned}
\lim_{x\to 0^+}\frac{\ln(1+x)}{\sin x} &= \lim_{x\to 0^+}\frac{\frac{1}{1+x}}{\cos x} \qquad \left(\text{form}\frac{0}{0}, \quad \text{Apply L'Hospitals rule}\right. \\[2mm]
&= \frac{1}{\cos 0} \\[2mm]
&= 1.
\end{aligned}
$$

(ii) Consider $\lim\limits_{x\to 0^+}\dfrac{\tan x - x}{x^3}$ which has the indeterminate form $\frac{0}{0}$.

$$
\begin{aligned}
\lim_{x\to 0^+}\frac{\tan x - x}{x^3} &= \lim_{x\to 0^+}\frac{\sec^2 x - 1}{3x^2} \qquad \left(\text{form}\frac{0}{0}, \text{Apply L'Hospitals rule}\right. \\[2mm]
&= \lim_{x\to 0^+}\frac{2\sec x \sec x \tan x}{6x} \\[2mm]
&= \lim_{x\to 0^+}\frac{2\frac{d}{dx}\left[\sec^2 x \tan x\right]}{6} \\[2mm]
&= \frac{1}{3}\lim_{x\to 0^+}\left[\sec^2 x \cdot \sec^2 x + 2\sec x \cdot \sec x \cdot \tan^2 x\right] \\[2mm]
&= \frac{1}{3}[1+0] \\[2mm]
&\; 1
\end{aligned}
$$

5. Evaluate the following limits:

(i) $\lim\limits_{x\to 0+} x^3 \ln x$ $(0,\infty)$,   (ii) $\lim\limits_{x\to\infty} \dfrac{x^3}{e^3}$ $(0,\infty)$.

Solution: (i) Consider $\lim\limits_{x\to 0+} x^3 \ln x$ which has the indeterminate form $0.(-\infty)$.

$$
\begin{aligned}
\lim_{x\to 0+} x^3 \ln x &= \lim_{x\to 0+} \frac{\ln x}{\frac{1}{x^3}} \qquad \left(\text{form}\frac{\infty}{\infty}, \ \text{Apply L'Hospitals rule}\right)\\
&= \lim_{x\to 0+} \frac{\frac{1}{x}}{\frac{-3}{x^4}} = \lim_{x\to 0+} \frac{x^3}{-3} = -\frac{1}{3}\lim_{x\to 0+} x^2 = \frac{-1}{3}\cdot 0 = 0.
\end{aligned}
$$

(ii) Consider $\lim\limits_{x\to\infty} \dfrac{x^3}{e^x}$ which has the indeterminate form $\frac{\infty}{\infty}$.

$$
\begin{aligned}
\lim_{x\to\infty} \frac{x^3}{e^x} &= \lim_{x\to\infty} \frac{3x^2}{e^x} \qquad \left(\text{form}\frac{\infty}{\infty}, \ \text{Apply L'Hospitals rule}\right)\\
&= \lim_{x\to\infty} \frac{6x}{e^x} \qquad \left(\text{form}\frac{\infty}{\infty}, \ \text{Apply L'Hospitals rule}\right)\\
&= \lim_{x\to\infty} \frac{6}{e^x} = 6\lim_{x\to\infty} \frac{1}{e^x}\\
&= 6\cdot 0 = 0.
\end{aligned}
$$

6. Evaluate the following limits:

(i) $\lim\limits_{x\to\infty} (1+3/x)^x$ $(0,\infty)$,   (ii) $\lim\limits_{x\to 0+} \left(\dfrac{1}{x} - \dfrac{1}{\arctan x}\right)$ $(0,\infty)$.

Solution: (i) Consider $\lim\limits_{x\to\infty} \left(1+\frac{3}{x}\right)^x$ $(1^\infty$ form$)$.

It is true that $\lim\limits_{x\to\infty} \left(1+\frac{1}{x}\right)^x = e$. It follows that in general $\lim\limits_{x\to\infty} \left(1+\frac{k}{x}\right)^x = e^k$.
Therefore

$$
\lim_{x\to\infty} \left(1+\frac{3}{x}\right)^x = e^3.
$$

(b) Consider

$$
\begin{aligned}
\lim_{x\to 0+} \left(\frac{1}{x} - \frac{1}{\arctan x}\right) &= \lim_{x\to 0+} \left(\frac{\arctan(x) - x}{x\arctan(x)}\right) \qquad \left(\text{form}\frac{0}{0}, \text{Apply L'Hospitals rule}\right.\\
&= \lim_{x\to 0+} \frac{\frac{1}{x^2+1} - 1}{\arctan x + \frac{x}{1+x^2}}\\
&= \lim_{x\to 0+} \frac{-x^2}{x^2\arctan x + x + \arctan x}\\
&= \lim_{x\to 0+} \frac{-2x}{2x\arctan x + 2}\\
&= \lim_{x\to 0+} \frac{-x}{x\arctan x + 1}\\
&= \frac{0}{1}\\
&= 0.
\end{aligned}
$$

7. Evaluate $\lim\limits_{x\to 0+} (\sin x)^x, \quad (0,\pi)$.

Solution: Now, $\lim\limits_{x\to 0+} (\sin x)^x, \quad$ is 0^0 form . By applying exponent rule, we have

$$\lim_{x\to 0+} (\sin x)^x = \lim_{x\to\infty} e^{x\ln\sin x}.$$

By the limit chain rule we know that if $\lim\limits_{u\to b} f(u) = L$ and $\lim\limits_{x\to a} g(x) = b$ and $f(x)$ is continuous at $x = b$ then $\lim\limits_{x\to a} f(g(x)) = L$.

For $u = g(x) = x\ln\sin x$ and $f(u) = e^u$. we have

$$\lim_{x\to 0+} x\ln\sin x = \lim_{x\to 0+} \frac{\ln\sin x}{\frac{1}{x}} \qquad (\text{form}\frac{\infty}{\infty}, \quad \text{Apply L'Hospitals rule})$$

$$= \lim_{x\to 0+} \frac{\cot x}{\frac{-1}{x^2}} = \lim_{x\to 0+} (-x^2\cot x)$$

$$= -\lim_{x\to 0+} (x^2\cot x) = -\lim_{x\to 0+} \frac{x^2}{\frac{1}{\cot x}}$$

$$= \lim_{x\to 0+} \frac{2x}{\sec^2 x}$$

$$= -\lim_{x\to 0+} 2x\cos^2 x$$

$$= -2\cdot 0\cdot 0 = 0.$$

$$\lim_{u\to 0+} e^u = \lim_{u\to 0+} e^0 = 1.$$

$$\therefore \lim_{x\to 0+} (\sin x)^x = 1.$$

8. Let f be differentiable on $(0,1)$ and suppose that $f(x) = \dfrac{e^x f(x)}{e^x}$, $\lim\limits_{x\to\infty} (f(x) + f'(x)) = L$. Show that $\lim\limits_{x\to\infty} f(x) = L$ and $\lim\limits_{x\to\infty} f'(x) = 0$.

Solution: We define a function f as

$$f(x) = \frac{e^x f(x)}{e^x}.$$

Now, the limit of f equal to

$$\lim_{x\to\infty} f(x) = \lim_{x\to\infty} \frac{e^x f(x)}{e^x}, \qquad \left(\frac{\infty}{\infty} \text{ form and by L'Hospital's rule}\right)$$

$$= \lim_{x\to\infty} \frac{e^x(f(x) + f'(x))}{e^x}$$

$$= \lim_{x\to\infty} \frac{e^x f(x) + e^x f'(x)}{e^x},$$

$$= \lim_{x\to\infty} \frac{e^x(f(x) + f'(x))}{e^x}$$

$$= \lim_{x\to\infty} (f(x) + f'(x))$$

$$= L.$$

We have shown that $\lim\limits_{\to\infty} f(x) = L$.

Now, lets see what is the limit of f' equal to:

From conditions given in exercise we know that

$$\lim_{x \to \infty} \left(f(x) + f'(x) \right) = L$$

It follows that

$$\lim_{x \to \infty} f(x) + \lim_{x \to \infty} f'(x) + L$$

Hence:

$$\lim_{x \to \infty} f'(x) = L - \lim_{x \to \infty} f'(x) = L - L = 0$$

We have shown that $\lim_{x \to \infty} f'(x) = 0$

9. Try to use L'Hospitals Rule to find the limit of $\dfrac{\tan x}{\sec x}$ as $x \to (\pi/2)$. Then evaluate directly by changing to sines and cosines.

   Solution: We can't evaluate the limit using L'Hosptial rule. If we tried we shall get

$$
\begin{aligned}
\lim_{x \to \frac{\pi}{2}} \frac{\tan x}{\sec x} &= \lim_{x \to \frac{\pi}{2}} \frac{\sec^2 x}{\sec x \tan x} && \left(\text{ form } \frac{\infty}{\infty}, L'H \right) \\
&= \lim_{x \to \frac{\pi}{2}} -\frac{\sec x}{\tan x} \\
&= \lim_{x \to \frac{\pi}{2}} \frac{\sec x \tan x}{\sec^2 x} && \left(\text{ form } \frac{\infty}{\infty}, L'H \right) \\
&= \lim_{x \to \frac{\pi}{2}} \frac{\tan x}{\sec x} \text{ is again of the form } \frac{\infty}{\infty} \cdots
\end{aligned}
$$

Using the definition of $\tan x$ and $\cos x$

$$
\begin{aligned}
\lim_{x \to \frac{\pi}{2}^-} \frac{\tan x}{\sec x} &= \lim_{x \to \frac{\pi}{2}^-} \frac{\frac{\sin x}{\cos x}}{\frac{1}{\cos x}} \\
&= \lim_{x \to \frac{\pi}{2}^-} \sin x \\
&= 1.
\end{aligned}
$$

2.3   Taylors Theorem

In this section, we discuss the Taylors theorem is a powerful result that has many applications.

Theorem 2.4. *Taylors Theorem*

Let $n \in \mathbb{N}$, let $I := [a, b]$, and let $f : I \to \mathbb{R}$ be such that f and its derivatives $f', f'', \cdots, f^{(n)}$ are continuous on I and that $f^{(n+1)}$ exists on (a, b). If $x_0 \in I$, then for any x in I there exists a point c between x and x_0 such that

$$f(x) = f(x_0) + f'(x_0)(x - x_0) + \frac{f''(x_0)}{2!}(x - x_0)^2 + \cdots + \frac{f^{(n)}(x_0)}{n!}(x - x_0)^n + R_n \qquad (2.1)$$

If we put $x - x_0 = h$ then the equation (2.2) becomes

$$f(x_0 + h) = f(x_0) + hf'(x_0) + \frac{h^2}{2!}\frac{f''(x_0)}{+} \cdots + \frac{h^2}{n!}\frac{f^{(n)}(x_0)}{n!} + R_n \tag{2.2}$$

where $R_n = \frac{h^{n+1}}{n+1!}f^{(n+1)}(c)$ is called Lagrange's form of remainder after n terms.

If $R_n \to 0$ as $n \to \infty$ and $x_0 = a$ then from equation (2.2), we obtain

$$f(x) = f(a) + f'(a)(x - a) + \frac{f''(a)}{2!}(x - a)^2 + \cdots + \frac{f^{(n)}(a)}{n!}(x - a)^n + \cdots \tag{2.3}$$

is called Taylor series of $f(x)$ about point a.

If we put $a = 0$ in equation (2.3) then the expansion

$$f(x) = f(0) + xf'(0) + \frac{x^2}{2!}f''(0) + \cdots + \frac{x^n}{n!}f^{(n)}(0) + \cdots \tag{2.4}$$

is called Maclaurin series.

<u>Illustrative Examples</u>

Example 2.5. *Find the Taylor series for following functions*

(i) e^x,   *(ii)* $\sin x$,   *(iii)* $\cos x$.

Solution: (i) Let $f(x) = e^x$.

$$f'(x) = e^x, \cdots, f^{(n)}(x) = e^x.$$

$$\therefore f'(0) = 1, \cdots, f^{(n)}(0) = 1.$$

Therefore, by Taylor series, we have

$$f(x) = f(0) + xf'(0) + \frac{x^2}{2!}f''(0) + \cdots + \frac{x^n}{n!}f^{(n)}(0) + \cdots \tag{2.5}$$

Putting above values in equation (2.6), we get

$$e^x = 1 + x + \frac{x^2}{2!} + \cdots + \frac{x^n}{n!} + \cdots$$

(ii) Let $f(x) = \sin x$.

Therefore, the derivatives of the function are

$$\begin{aligned}
f(x) &= \sin x & f(0) &= 0 \\
f'(x) &= \cos x & f'(0) &= 1 \\
f''(x) &= -\sin x & f''(0) &= 0 \\
f'''(x) &= -\cos x & f'''(0) &= -1 \\
&\;\;\vdots & &\;\;\vdots
\end{aligned}$$

Hence, we can conclude the $f^{(n)}(0) = 0$ if n is even and $f^{(n)}(0) = \pm 1$ if n is odd.

Therefore, by Taylor series, we have

$$f(x) = f(0) + x f'(0) + \frac{x^2}{2!} f''(0) + \cdots + \frac{x^n}{n!} f^{(n)}(0) + \cdots \tag{2.6}$$

Putting above values in equation (2.6), we get

$$\sin x = x - \frac{x^3}{3!} + \frac{x^5}{5!} - \cdots .$$

(iii) Let $f(x) = \cos x$.

Therefore, the derivatives of the function are

$$\begin{aligned}
f(x) &= \cos x &\qquad f(0) &= 1 \\
f'(x) &= -\sin x &\qquad f'(0) &= 0 \\
f''(x) &= -\cos x &\qquad f''(0) &= -1 \\
f'''(x) &= \sin x &\qquad f'''(0) &= 0 \\
f^{iv}(x) &= \cos x &\qquad f^{iv}(0) &= 1 \\
&\ \ \vdots &\qquad &\ \ \vdots
\end{aligned}$$

Hence, we can conclude the $f^{(n)}(0) = \pm 1$ if n is even and $f^{(n)}(0) = 0$ if n is odd.

Therefore, by Taylor series, we have

$$f(x) = f(0) + x f'(0) + \frac{x^2}{2!} f''(0) + \cdots + \frac{x^n}{n!} f^{(n)}(0) + \cdots \tag{2.7}$$

Putting above values in equation (2.7), we get

$$\cos x = 1 - \frac{x^2}{2!} + \frac{x^4}{4!} - \cdots .$$

Example 2.6. *Find the Taylor series for $f(x) = \sqrt{1 + x}$.*

Solution: Let $f(x) = \sqrt{1 + x}$.

Therefore, the derivatives of the function are

$$\begin{aligned}
f(x) &= \sqrt{1 + x} &\qquad f(0) &= 1 \\
f'(x) &= \frac{1}{\sqrt{1 + x}} &\qquad f'(0) &= \frac{1}{2} \\
f''(x) &= \frac{1}{(1 + x)^{1/2}} &\qquad f''(0) &= \frac{-1}{4} \\
&\ \ \vdots &\qquad &\ \ \vdots
\end{aligned}$$

Therefore, by Taylor series, we have

$$x^2 \qquad\qquad\qquad x^n$$

Putting above values in equation (2.8), we get

$$f(x) = 1 + \frac{x}{2} - \frac{x^2}{8} + \cdots .$$

Example 2.7. *Find the first three terms of the Taylor series for the function* $\sin(\pi x)$ *centered at* $a = \frac{1}{2}$. *Use your answer to find an approximate value to* $\sin(\frac{\pi}{2} + \frac{\pi}{10})$.

Solution: Let $f(x) = \sin(\pi x)$ and $a = \frac{1}{2}$.

Therefore, the derivatives of the function are

$$f(x) = \sin(\pi x) \qquad\qquad f(\tfrac{1}{2}) = 1$$

$$f'(x) = \pi \cos(\pi x) \qquad\qquad f'(\tfrac{1}{2}) = 0$$

$$f''(x) = -\pi^2 \sin(\pi x) \qquad\qquad f''(\tfrac{1}{2}) = -\pi^2$$

$$f'''(x) = -\pi^3 \cos(\pi x) \qquad\qquad f'''(\tfrac{1}{2}) = 0$$

$$f''''(x) = \pi^4 \sin(\pi x) \qquad\qquad f''''(\tfrac{1}{2}) = \pi^4$$

$$\vdots \qquad\qquad\qquad\qquad \vdots$$

Therefore, by Taylor series, we have

$$f(x) = f(a) + f'(a)(x-a) + \frac{f''(a)}{2!}(x-a)^2 + \cdots + \frac{f^{(n)}(a)}{n!}(x-a)^n + \cdots$$

$$f(x) = f(\tfrac{1}{2}) + f'(\tfrac{1}{2})(x - \tfrac{1}{2}) + \frac{f''(\tfrac{1}{2})}{2!}(x - \tfrac{1}{2})^2 + \frac{f'''(\tfrac{1}{2})}{3!}(x - \tfrac{1}{2})^3 + \frac{f''(\tfrac{1}{2})}{4!}(x - \tfrac{1}{2})^4 + \cdots \qquad (2.9)$$

Putting above values in equation (2.9), we get

$$\sin(\pi x) = 1 + 0(x - \tfrac{1}{2}) + \frac{-\pi^2}{2!}(x - \tfrac{1}{2})^2 + \frac{0}{3!}(x - \tfrac{1}{2})^3 + \frac{\pi^4}{4!}(x - \tfrac{1}{2})^4 + \cdots$$

$$\sin(\pi x) = 1 - \frac{\pi^2}{2!}(x - \tfrac{1}{2})^2 + \frac{\pi^4}{4!}(x - \tfrac{1}{2})^4 + \cdots$$

$$\therefore \sin(\frac{\pi}{2} + \frac{\pi}{10}) = 1 - \frac{\pi^2}{2!}(\frac{1}{10})^2 + \frac{\pi^4}{4!}(\frac{1}{10})^4$$

$$= 1 - 0.0493 + 0.0004$$

$$= 0.9511.$$

Example 2.8. *Using Taylor series expansion find the approximate value of* $\sqrt{25.15}$.

Solution: Let $f(x) = \sqrt{x}$, $a = 25$, $h = 0.15$.

Therefore, the derivatives of the function are

$$
\begin{aligned}
f(x) &= \sqrt{x} & f(25) &= 5 \\
f'(x) &= \frac{1}{2\sqrt{x}} & f'(25) &= 0.1 \\
f''(x) &= -\frac{1}{4x^{\frac{3}{2}}} & f''(25) &= 0.002 \\
&\ \vdots & &\ \vdots
\end{aligned}
$$

Therefore, by Taylor series, we have

$$
f(a+h) = f(a) + hf'(a) + h^2\frac{f''(a)}{2!} + \cdots + h^n\frac{f^{(n)}(a)}{n!} + \cdots
$$

Putting above values, we get

$$
f(25 + 0.15) = f(25) + 0.15f'(25) + 0.15^2\frac{f''(25)}{2!} + \cdots
$$

$$
f(25.15) \approx 5 + 0.15 \times 0.1 + 0.0225 \times \frac{1}{2!} \times 0.002 + \cdots
$$

$$
\sqrt{25.15} \approx 5.0149775
$$

Example 2.9. *Find the Maclaurin series for* $x \sin x$.

Solution: We know that

$$
\sin x = x - \frac{x^3}{3!} + \frac{x^5}{5!} - \cdots .
$$

Therefore, multiplying by x, we obtain the Maclaurin series for $x\sin x$ as follows

$$
x\sin x = x^2 - \frac{x^4}{3!} + \frac{x^6}{5!} - \cdots .
$$

Example 2.10. *To approximate sine by a polynomial on* $[-1,1]$ *so that the error is less than* 0.001. *Show that*

$$
\left| \sin x - \left(x - \frac{x^3}{6} + \frac{x^5}{120} \right) \right| < \frac{1}{5040} \text{ for } |x| \le 1.
$$

Solution: Let $f(x) = \sin x$ for $x \in [-1,1]$, then:

$$
\begin{aligned}
f'(x) &= \cos x & f'(0) &= 1 \\
f''(x) &= -\sin x & f''(0) &= 0 \\
f^{(3)}(x) &= -\cos x & f^{(3)}(0) &= -1 \\
f^{(4)}(x) &= \sin x & f^{(4)}(0) &= 0 \\
f^{(5)}(x) &= \cos x & f^{(5)}(0) &= 1 \\
f^{(6)}(x) &= -\sin x & f^{(6)}(0) &= 0 \\
f^{(7)}(x) &= -\cos x & f^{(7)}(0) &= -1 \\
&\ \vdots
\end{aligned}
$$

Using Taylor's theorem, we get

$$
f(x) = f(0) + f'(0) + \frac{x^2}{2!}f''(0) + \frac{x^3}{3!}f^{(3)}(0) + \frac{x^4}{4!}f^{(4)}(0) + \frac{x^5}{5!}f^{(5)}(0)
$$

$$
+ \frac{x^6}{\,}f^{(6)}(0) + \frac{x^7}{\,}f^{(7)}(\ )
$$

where $c \in (x, 0)$, we obtain

$$\sin x = x - \frac{x^3}{3!} + \frac{x^5}{5!} + \frac{x^7}{7!} \cdot (-\cos c)$$

$$\left| \sin x - \left(x - \frac{x^3}{3!} + \frac{x^5}{5!} \right) \right| = \left| \frac{x^7}{7!} \cdot (-\cos c) \right|.$$

As $|\cos x| \leq 1$ and $|\cos c| \leq 1$ it follows that

$$\left| \sin x - \left(x - \frac{x^3}{3!} + \frac{x^5}{5!} \right) \right| = \left| \frac{x^7}{7!} \cdot (-\cos c) \right| \leq \frac{|x|^7}{7!} \leq \frac{1}{7!}$$

Therefore, for $|x| \leq 1$ we proved

$$\left| \sin x - \left(x - \frac{x^3}{3!} + \frac{x^5}{5!} \right) \right| \leq \frac{1}{5040}.$$

2.4   Successive Differentiation

Suppose f be a differentiable function of x then we write $\dfrac{dy}{dx} = f'(x)$ is called first derivative of f with respect to x. Again, if the derivative $f'(x)$ is a function of x and if $f'(x)$ is differentiable at x, then the derivative of f' at x is called second derivative of f at x and it is denoted by $\dfrac{d^2y}{dx^2} = f''(x)$. Similarly, if $''$ is differentiable at x , then this derivative is called the third derivative of f and it is denoted by $\dfrac{d^3y}{dx^3} = f'''(x)$. Proceeding in this way the nth derivative of f is the derivative of the function $f^{(n-1)}(x)$ and it the denoted by $\dfrac{d^ny}{dx^n} = f^{(n)}(x)$.

The expressions

$$\frac{dy}{dx}, \frac{d^2y}{dx^2}, \frac{d^3y}{dx^3}, \cdots, \frac{d^ny}{dx^n}$$

are called respectively first, second, third and so on n^{th} order derivatives of the function f w.r.t. x. These derivatives are also, denoted by

$$f'(x), f''(x), f'''(x), \cdots, f^{(n)}(x).$$

$$y', y'', y''', \cdots, y^{(n)} \quad \text{or} \quad y_1, y_2, y_3, \cdots, y_n.$$

2.4.1   The nth Derivatives of standard Functions

In this section, we obtain the nth derivatives of standard functions and solved some examples.

(i) Let $y = (ax + b)^m$.

   Therefore,

$$y_1 = ma(ax + b)^{m-1}$$

$$y_2 = m(m - 1)a^2(ax + b)^{m-2}$$

$$y_3 = m(m - 1)(m - 2)a^3(ax + b)^{m-3}$$

$$\vdots$$

$$y_n = a^n m(m-1)(m-2)\cdots(m-n+1)(ax+b)^{m-n}.$$

We can write this as

$$y_n = a^n m(m-1)(m-2)\cdots(m-n+1)(ax+b)^{m-n}.$$

$$y_n = \frac{m!}{(m-n)!}a^n(ax+b)^{m-n}.$$

If $m = n$ then $y_n = n!a^n$.

If $m = -1$ then

$$y_n = (-1)(-2)(-3)\cdots(-n)a^n(ax+b)^{-1-n}$$

$$= (-1)^n 1.2.3.\cdots na^n(ax+b)^{-1-n}$$

$$= (-1)^n 1.2.3.\cdots na^n(ax+b)^{-1-n}$$

$$y_n = (-1)^n.n!.a^n(ax+b)^{-1-n}$$

$$D^n\left(\frac{1}{(ax+b)}\right) = \frac{(-1)^n n! a^n}{(ax+b)^{n+1}}. \qquad (2.10)$$

(ii) Let $y = log(ax+b)$.

Therefore,

$$y_1 = \frac{a}{(ax+b)}$$

Use the formula (2.10), for $n-1$, we get

$$y_n = \frac{(-1)^{n-1}(n-1)!a^n}{(ax+b)^n}.$$

(iii) Let $y = a^{mx}$. Therefore,

$$y_1 = ma^{mx}(\log a), \;\; y_2 = m^2 a^{mx}(\log a)^2$$

$$y_3 = m^3 a^{mx}(\log a)^3, \cdots$$

$$y_n = m^n a^{mx}(\log a)^m.$$

Put $a = e$, we get

$$y_n = m^n e^{mx}.$$

(iv) Let $y = \sin(ax+b)$. Therefore,

$$y_1 = a\cos(ax+b) = a\sin\left[(ax+b) + \frac{\pi}{2}\right]$$

$$y_2 = -a^2\sin(ax+b) = a^2\sin\left[(ax+b) + \frac{2\pi}{2}\right]$$

$$y_3 = -a^3 \cos(ax + b) = a^3 \sin\left[(ax + b) + \frac{3\pi}{2}\right]$$

$$\vdots$$

$$y_n = a^n \sin[(ax + b) + \frac{n\pi}{2}].$$

Similarly, if $y = \cos(ax + b)$ then

$$y_n = a^n \cos\left[(ax + b) + \frac{n\pi}{2}\right].$$

(v) Let $y = e^{ax} \sin(bx + c)$. Therefore,

$$y_1 = ae^{ax} \sin(bx + c) + be^{ax} \cos(bx + c)$$
$$= e^{ax}[a \sin(bx + c) + b \cos(bx + c)].$$

Putting, $a = r \cos \theta$ and $b = r \sin \theta$, we get

$$r = \sqrt{a^2 + b^2}, \ \theta = tan^{-1}\left(\frac{b}{a}\right).$$

$$y_1 = e^{ax}[r \cos \theta \sin(bx + c) + r \sin \theta \cos(bx + c)]$$
$$= re^{ax} \sin(bx + c + \theta)$$
$$y_2 = r^2 e^{ax} \sin(bx + c + 2\theta)$$

$$\vdots$$

$$y_n = r^n e^{ax} sin(bx + c + n\theta).$$

Similarly, if $y = e^{ax} \cos(bx + c)$ then $y_n = r^n e^{ax} \cos(bx + c + n\theta)$.

Example 2.11. *Find the nth derivatives of $\dfrac{x^4}{(x-1)(x-2)}$.*

Solution: Let $y = \dfrac{x^4}{(x-1)(x-2)}$.
Therefore,

$$\frac{x^4}{(x-1)(x-2)} = \frac{A}{(x-1)} + \frac{B}{(x-2)}$$
$$x^4 = A(x-2) + B(x-2)$$

Putting, $x = 1, 2$, we obtain $A = -1, B = 16$.

$$\frac{x^4}{(x-1)(x-2)} = \frac{-1}{(x-1)} + \frac{16}{(x-2)}$$

$$\frac{x^4}{(x-1)(x-2)} = \frac{16}{(x-2)} - \frac{1}{(x-1)}.$$

Using formula $D^n\left(\dfrac{1}{(ax+b)}\right) = \dfrac{(-1)^n n! a^n}{(ax+b)^{n+1}}$, we get

$$
\begin{aligned}
y_n &= \frac{d^n}{dx^n}\left[\frac{x^4}{(x-1)(x-2)}\right] \\
&= \frac{d^n}{dx^n}\left[\frac{16}{(x-2)} - \frac{1}{(x-1)}\right] \\
&= 16\frac{d^n}{dx^n}\left[\frac{1}{(x-2)}\right] - \frac{d^n}{dx^n}\left[\frac{1}{(x-1)}\right] \\
&= 16\frac{(-1)^n n! 1^n}{(x-2)^{n+1}} - \frac{(-1)^n n! 1^n}{(x-1)^{n+1}} \\
\therefore\ y_n &= (-1)^n n!\left[\frac{16}{(x-2)^{n+1}} - \frac{1}{(x-1)^{n+1}}\right].
\end{aligned}
$$

Example 2.12. *Find the nth derivatives of $\cos^4 x$.*

Solution: Let $y = \cos^4 x$. We have

$$
\begin{aligned}
\cos^2 x &= \frac{1+\cos 2x}{2} \\
\cos^4 x &= \left(\frac{1+\cos 2x}{2}\right)^2 \\
&= \frac{1}{4} + \frac{\cos 2x}{2} + \frac{\cos^2 2x}{4} \\
&= \frac{1}{4} + \frac{\cos 2x}{2} + \frac{1+\cos 4x}{8}.
\end{aligned}
$$

$$
\begin{aligned}
\frac{d^n}{dx^n}(\cos^4 x) &= \frac{d^n}{dx^n}\left[\frac{1}{4} + \frac{\cos 2x}{2} + \frac{1+\cos 4x}{8}\right] \\
&= \frac{d^n}{dx^n}\left(\frac{1}{4}\right) + \frac{1}{2}\frac{d^n}{dx^n}(\cos 2x) + \frac{1}{8}\frac{d^n}{dx^n}(1+\cos 4x) \\
&= \frac{1}{2}2^n \cos\left(2x+\frac{n\pi}{2}\right) + \frac{1}{8}4^n \cos\left(4x+\frac{n\pi}{2}\right) \\
&= 2^{n-1}\cos\left(2x+\frac{n\pi}{2}\right) + 2^{2n-3}\cos\left(4x+\frac{n\pi}{2}\right).
\end{aligned}
$$

Example 2.13. *Find the nth derivatives of $e^{2x}\cos^2 x \sin x$.*

Solution: Let $y = e^{2x}\cos^2 x \sin x$. We have

$$
\begin{aligned}
\cos^2 x \sin x &= \frac{1+\cos 2x}{2}\sin x \\
&= \frac{1}{2}\sin x + \frac{1}{2}\sin x \cos 2x \\
&= \frac{1}{2}\sin x + \frac{1}{4}(\sin 3x \sin x) \\
&= \frac{1}{4}\sin x + \frac{1}{4}\sin 3x.
\end{aligned}
$$

$$\frac{d^n}{dx^n}(e^{2x}\cos^2 x \sin x) = \frac{d^n}{dx^n}\left[\frac{1}{4}e^{2x}\sin x + \frac{1}{4}e^{2x}\sin 3x\right]$$

$$= \frac{1}{4}\frac{d^n}{dx^n}(e^{2x}\sin x) + \frac{1}{4}\frac{d^n}{dx^n}(e^{2x}\sin 3x)$$

$$= \frac{1}{4}5^{n/2}\sin\left(x + n\tan^{-1}\left(\frac{1}{2}\right)\right) + \frac{1}{4}13^{n/2}\sin\left(3x + n\tan^{-1}\left(\frac{3}{2}\right)\right).$$

2.4.2 Leibnitz's Theorem

Leibnitz's theorem is useful to obtain n^{th} derivative of the product of two functions.

Theorem 2.5. *If u, v be two functions possessing derivatives of the nth order, then*

$$(uv)_n = {}^nC_0 u_n v + {}^nC_1 u_{n-1}v_1 + {}^nC_2 u_{n-2}v_2 + \cdots + {}^nC_r u_{n-r}v_r + \cdots + {}^nC_n uv_n$$

where

$$ {}^nC_r = \frac{n!}{(n-r)!}, \quad {}^nC_{r-1} + {}^nC_r = {}^{n+1}C_r \text{ and } {}^nC_n = {}^{n+1}C_{n+1} = 1.$$

Proof: We prove this theorem by Mathematical induction. We know that

$$(uv)_1 = u_1 v + uv_1$$

$$= {}^1C_0 u_1 v + {}^1C_1 uv_1$$

$$(uv)_2 = u_2 v + u_1 v_1 + u_1 v_1 + uv_2$$

$$= {}^2C_0 u_2 v + {}^2C_1 u_1 v_1 + {}^2C_2 uv_2.$$

Thus result is true for $n = 1, 2$.

Suppose that result is true for $n = m$.

$$(uv)_m = {}^mC_0 u_m v + {}^mC_1 u_{m-1}v_1 + {}^mC_2 u_{m-2}v_2 + \cdots + {}^mC_r u_{m-r}v_r + \cdots + {}^mC_m uv_m. \qquad (2.11)$$

We have to prove result is true for $n = m + 1$.

Differentiating equation (2.11) w.r.t. x, we get

$$(uv)_{m+1} = {}^mC_0 u_{m+1}v + {}^mC_0 u_m v_1 + {}^mC_1 u_m v_1 + {}^mC_1 u_{m-1}v_2 + {}^mC_2 u_{m-1}v_2 + {}^mC_2 u_{m-2}v_3 + \cdots$$

$$+ {}^mC_{r-1}u_{m-r+2}v_{r-1} + {}^mC_{r-1}u_{m-r+1}v_r + {}^mC_r u_{m-r+1}v_r + {}^mC_r u_{m-r}v_{r+1} + \cdots + {}^mC_m uv_{m+1}$$

$$= {}^mC_0 u_{m+1}v + ({}^mC_0 + {}^mC_1)u_m v_1 + ({}^mC_1 + {}^mC_2)u_{m-1}v_2 + \cdots + ({}^mC_{r-1} + {}^mC_r)u_{m-r+1}v_r +$$

$$\cdots + {}^mC_m uv_{m+1}$$

$$= {}^{m+1}C_0 u_{m+1}v + {}^{m+1}C_1 u_m v_1 + {}^{m+1}C_2 u_{m-1}v_2 + \cdots + {}^{m+1}C_r u_{m-r+1}v_r + \cdots + {}^{m+1}C_{m+1}uv_{m+1}.$$

Therefore, result is true for $n = m + 1$.

Hence, by induction, we prove the result is true for all n.

where

$$^nC_r = \frac{n!}{(n-r)!}, \quad ^nC_{r-1} + {}^nC_r = {}^{n+1}C_r \text{ and } {}^nC_n = {}^{n+1}C_{n+1} = 1.$$

Example 2.14. *Find the nth derivatives of $x^2 e^x \cos x$.*

Solution: Let $y = x^2 e^x \cos x$.

Suppose $u = e^x \cos x$ and $v = x^2$. Applying Leibnitz's theorem, we get

$$\frac{d^n}{dx^n}(x^2 e^x \cos x) = (x^2 e^x \cos x)_n$$

$$= (e^x \cos x)_n . x^2 + {}^nC_1 (e^x \cos x)_{n-1} . 2x + {}^nC_2 (e^x \cos x)_{n-2} . 2$$

$$= 2^{n/2} e^x \cos(x + n \tan^{-1}(1)) x^2 + n 2^{(n-1)/2} e^x \cos(x + (n-1)\tan^{-1}(1)) 2x$$

$$+ \frac{n(n-1)}{2!} 2^{(n-2)/2} e^x \cos(x + (n-2)\tan^{-1}(1)) 2$$

$$= 2^{(n-2)/2} e^x \left[2x^2 \cos\left(x + \frac{n\pi}{4}\right) + 2^{3/2} nx \cos\left(x + \frac{(n-1)\pi}{4}\right) \right.$$

$$\left. + n(n-1) \cos\left(x + \frac{(n-2)\pi}{4}\right) \right].$$

Example 2.15. *If $y = a\cos(\log x) + b\sin(\log x)$, prove that $x^2 y_{n+2} + (2n+1)xy_{n+1} + (n^2+1)y_n = 0$.*

Solution: Let $y = a\cos(\log x) + b\sin(\log x)$.

Differentiating w.r.t. x, we get

$$y_1 = \frac{-a\sin(\log x)}{x} + \frac{b\cos(\log x)}{x}$$

$$xy_1 = -a\sin(\log x) + b\cos(\log x)$$

$$xy_2 + y_1 = \frac{-a\cos(\log x)}{x} - \frac{b\sin(\log x)}{x}$$

$$x^2 y_2 + xy_1 = -[a\cos(\log x) + b\sin(\log x)]$$

$$x^2 y_2 + xy_1 = -y$$

$$x^2 y_2 + xy_1 + y = 0.$$

Applying Leibnitz's theorem, we get

$$x^2 y_{n+2} + {}^nC_1 y_{n+1} . 2x + {}^nC_2 y_n . 2 + xy_{n+1} + {}^nC_1 y_n . 1 + y_n = 0$$

$$x^2 y_{n+2} + 2nxy_{n+1} + n(n-1)y_n + xy_{n+1} + ny_n . 1 + y_n = 0$$

$$x^2 y_{n+2} + (2n+1)xy_{n+1} + (n^2+1)y_n = 0.$$

Example 2.16. *If $y^{\frac{1}{m}} + y^{\frac{-1}{m}} = 2x$, prove that $(x^2-1)y_{n+2} + (2n+1)xy_{n+1} + (n^2 - m^2)y_n = 0$.*

Solution: Let $y^{\frac{1}{m}} + y^{\frac{-1}{m}} = 2x$.

Differentiating w.r.t. x, we get

$$\frac{y^{\frac{1}{m}}-1}{m}y_1 - \frac{y^{\frac{-1}{m}}-1}{m}y_1 = 2$$

$$y^{\frac{1}{m}} - y^{\frac{-1}{m}} = \frac{2my}{y_1}$$

$$(y^{\frac{1}{m}} - y^{\frac{-1}{m}})^2 = (y^{\frac{1}{m}} + y^{\frac{-1}{m}})^2 - 4$$

$$\left(\frac{2my}{y_1}\right)^2 = (2x)^2 - 4$$

$$\frac{4m^2y^2}{y_1^2} = 4x^2 - 4$$

$$(x^2 - 1)y_1^2 = m^2y^2.$$

Differentiating w.r.t. x, we get

$$(x^2 - 1)2y_1y_2 + 2xy_1^2 = 2m^2yy_1$$

$$(x^2 - 1)y_2 + xy_1 = m^2y.$$

Applying Leibnitz's theorem, we get

$$(x^2 - 1)y_{n+2} + {}^nC_1y_{n+1}.2x + {}^nC_2y_n.2 + xy_{n+1} + {}^nC_1y_n.1 = m^2y_n$$

$$(x^2 - 1)y_{n+2} + 2nxy_{n+1} + n(n-1)y_n + xy_{n+1} + ny_n - m^2y_n = 0$$

$$(x^2 - 1)y_{n+2} + (2n+1)xy_{n+1} + (n^2 - m^2)y_n = 0.$$

Exercise

1. Evaluate the following limits.

   (i) $\lim\limits_{x\to 0}\dfrac{e^x + e^{-x} - 2}{1 - \cos x}$ (ii) $\lim\limits_{x\to 0}\dfrac{x^2 - \sin^2 x}{x^4}$

2. Evaluate the following limits.

   (i) $\lim\limits_{x\to 0+}\dfrac{\tan x}{x}$ $(0, \pi/2)$, (ii) $\lim\limits_{x\to 0+}\dfrac{\ln\cos x}{x}$.

3. Evaluate the following limits.

   (i) $\lim\limits_{x\to 0}\dfrac{\arctan x}{x}$ $(-\infty, \infty)$, (ii) $\lim\limits_{x\to 0}\dfrac{1}{x(\ln x)^2}$ $(0, 1)$,.

4. Evaluate the following limits.

   (i) $\lim\limits_{x\to\infty}\dfrac{\ln x}{x^2}$ $(0, \infty)$,   (ii) $\lim\limits_{x\to\infty}\dfrac{\ln x}{\sqrt{x}}$ $(0, \infty)$,

   (iii) $\lim\limits_{x\to 0} x\ln\sin x$ $(0, \pi)$,   (iv) $\lim\limits_{x\to\infty}\dfrac{x + \ln x}{x\ln x}$ $(0, \infty)$.

5. Evaluate the following limits.

$$\left(\frac{3}{}\right)^x$$

6. Evaluate the following limits:

   (i) $\lim\limits_{x \to \infty} x^{1/x}$ $(0, \infty)$,   (ii) $\lim\limits_{x \to \pi/2-} (\sec x - \tan x)$ $(0, \pi/2)$.

7. Find the Taylor series for $f(x) = (x - 1)e^x$ about $x = 1$.

8. Obtain the first three terms and find the Maclaurin series for the following functions

   (i) $\sin^2 x$,   (ii) $\dfrac{x}{\sqrt{1 - x^2}}$.

9. Find the Maclaurin series for $\ln(1 + x)$ and hence that for $\ln\left(\dfrac{1 + x}{1 - x}\right)$.

10. Find the Taylor series for $f(x) = \dfrac{2}{(1 + x)^3}$ near $x = 1$.

11. Find the Taylor series for $f(x) = \dfrac{1}{x}$ centered at $a = 1$ and $f(x) = \cos x$ centered at $a = \frac{\pi}{2}$.

12. What is the coefficient of the x^5 term in the Maclaurin polynomial for $\sin 2x$.

13. Find a series formula for π by using Taylor series and $\tan^{-1}\left(\frac{1}{\sqrt{3}}\right) = \frac{\pi}{6}$.

14. Find the nth derivatives of $\dfrac{1}{(2x - 1)(3x - 1)}$.

15. Find the nth derivatives of the following functions

   (i) $\dfrac{4x}{(x - 1)^2(x + 1)}$,   (ii) $\dfrac{x + 1}{x^2 - 4}$,   (iii) $x^n e^x$,   (iv) $e^x \log x$,   (v) $e^{2x} \cos x \sin^2 2x$.

16. Find the nth derivatives of the following functions

   (i) $e^x \sin^4 x$,   (ii) $\sin^3 x$,   (iii) $x^2 \sin x$.

17. If $y = e^{m \sin^{-1} x}$, prove that $(1 - x^2)y_{n+2} - (2n + 1)xy_{n+1} - (n^2 + m^2)y_n = 0$.

18. If $y = (\sin^{-1} x)^2$, prove that $(1 - x^2)y_{n+2} - (2n + 1)xy_{n+1} - n^2 y_n = 0$.

19. If $y = \sin(m \sin^{-1} x)$, prove that $(1 - x^2)y_{n+2} - (2n + 1)xy_{n+1} + (n^2 - m^2)y_n = 0$.

20. If $y = e^{m \cos^{-1} x}$, prove that $(1 - x^2)y_{n+2} - (2n + 1)xy_{n+1} - (n^2 + m^2)y_n = 0$.

21. Find the Taylor series expansion for $f(x + h)$ in powers of x.

22. Expand $\log \cos(x + \frac{\pi}{4})$, using Taylor's theorem in ascending powers of x and hence find the value of $\log \cos(48)$.

23. Using Taylor series prove that $\log \tan(x + \frac{\pi}{4}) = 2x + \frac{4}{3}x^3 + \frac{4}{5}x^5 + \cdots$.

24. Using Taylor series expansion find the value of $(i)\sqrt{25.15}$, $(ii)\sqrt{37}$, $(iii)\sqrt{9.12}$.

25. Using Taylor series prove that $\log \sin(x + h) = \log(\sin x) + h \cot x - \dfrac{h^2}{2}\dfrac{1}{\cos^2 x} + \dfrac{h^3}{3}\dfrac{\cos x}{\sin^3 x} + \cdots$.

   Using Taylor series expand ^{3} ^{2} 6 in powers of $($ 3$)$

27. Using Taylor series expand e^x in powers of $(x-2)$.

28. Using Taylor series prove that $\tan^{-1} x = \frac{\pi}{4} + \frac{1}{2}(x-1) - \frac{1}{4}(x-1)^2 + \frac{1}{12}(x-1)^3 + \cdots$.

29. Using Maclaurin series prove that $\tan^{-1} x = x - \frac{x^3}{3} + \frac{x^5}{5} - \frac{x^7}{7} + \cdots$.

30. Using Maclaurin series prove that $\tan^{-1}\left(\frac{1-x}{1+x}\right) = \frac{\pi}{4} - \frac{1}{2}\left(x - \frac{x^3}{3} + \frac{x^5}{5} - \frac{x^7}{7}\right) + \cdots$.

31. Using Maclaurin series prove that $e^x \sin x = x + x^2 + \frac{2x^3}{3!} - \frac{4x^5}{5!} + \cdots$.

32. Using Maclaurin series prove that $e^{\sin x} = 1 + x + \frac{1}{2}x^2 + \cdots$.

33. Using Maclaurin series prove that $\log(1 + \sin x) = x - \frac{1}{2}x^2 + \frac{1}{3}x^3 - \frac{1}{12}x^4 + \cdots$.

Solutions

1. (i) 2, (ii) $\frac{1}{3}$.

2. (i) 1, (ii) 0.

3. (i) 1

4. (i) 0, (ii) 0 (iii) 0, (iv) 0.

5. (i) 1, (ii) e^3.

6. (i) 1, (ii) 0.

7. $e\left[(x-1) + (x-1)^2 + \frac{(x-1)^3}{2} + \frac{(x-1)^4}{6} + \cdots\right]$

8. (i) $x^2 - \frac{x^4}{3} + \frac{2x^6}{45} + \cdots\right]$

  (ii) $-3(1-x^2)^{-5/2} + 15(1-x^2)^{-5/2} + \cdots\right]$

10. $\frac{1}{4} - \frac{3}{8}(x-1) + \frac{3}{8}(x-1)^2 - \frac{5}{10}(x-1)^3 + \cdots\right]$

12. 0.26667

14. $y_n = (-1)^n n!\left[\frac{2^{n+1}}{(2x-1)^{n+1}} + \frac{3^{n+1}}{(3x-1)^{n+1}}\right]$.

44. (i) 5.0149775, (ii) 6.0827625,   (iii) 3.0199377.

26. $87 + 70(x-3) + 16(x-3)^2 + (x-3)^3$.

27. $e^2\left[1 + (x-2) + \frac{(x-2)^2}{2!} + \frac{(x-2)^3}{3!} + \cdots\right]$

Maxima Lab Work

1. Does the following functions are differentiable? Justify your answer by plotting graph.

    (i) $|x - 1|$, (ii) $x^{\frac{1}{3}}$.

2. Using limit definition of derivative, find the derivative of following functions.

    (i) $f(x) = 2x^3 + 3x + 1$ for $x \in \mathbb{R}$,   (ii) $g(x) = \dfrac{1}{\sqrt{2 - x}}$ for $x \in \mathbb{R}, x \neq 2$.

    (iii) $f(x) = x^3$ for $x \in \mathbb{R}$,   (iv) $g(x) = \dfrac{1}{x}$ for $x \in \mathbb{R}, x \neq 0$.

    (v) $h(x) = \sqrt{x}$ for $x > 0$,   (vi) $k(x) = \dfrac{1}{\sqrt{x}}$ for $x > 0$.

3. Find the derivative of a function using the diff command

    (i) $\cos x$, $\sin x$, $\cos^{-1} x$, $\sinh x$, $\cosh x$.

    (ii) Let $f(x) = \tan^{-1} x$, compute $f'(1.5)$ and find all x such that $f'(x) = \frac{1}{2}$. Also, plot $f(x)$ together with the tangent lines at $= \pm 1$.

4. Let $f(x) = 2 \sin x$. Use Maxima to compute $f'(x)$. Compute the equation of the tangent line to $f(x)$ at $x = 2$. Also, plot $f(x)$, $f'(x)$ and the tangent line at $x = 2$. Comment on the slope relationship between $f(x)$ and $f'(x)$ at $x = 2$.

5. Verify the quotient rule for the functions $f(x) = \sin x$ and $g(x) = \cos x$.

6. Find the derivative of $h(x) = e^{x^2}$ by chain rule in Maxima.

7. Let $f(x) = sqrt(\sin^2 x + 2)$. Find $f'(x)$ by chain rule in Maxima.

8. Use implicit differentiation to find the slope of a tangent line to the unit circle at any point (as a function of x and y). Use your answer to plot the tangent line at the point $\left(-\dfrac{1}{2}, \dfrac{\sqrt{3}}{2} \right)$.

9. Calculate the derivative of the following functions.

    (i) $(x^4 - 3x^2 + 5x - 2)^3$, (ii) $\sqrt{7x^3 - 2x^2 + 5}$, (iii) $\dfrac{1}{(5x^2 + 4)^3}$.

10. Let $f(x) = \tan x$, use the limit definition of the derivative to compute $f'(1)$. Also, plot a function $f(x)$ together with the tangent line at $(1, f(1))$.

11. For the function $f(x) = 0.5x^2 - 3x$, find the equation of the secant line connecting $(1, f(1))$ to $(3, f(3))$. Make a plot of $f(x)$ together with a plot of the secant line.

12. Verify Rolle's theorem for $f(x) = x^2 + 2x$ on $[-2, 0]$.

13. Verify the Mean Value theorem for the function $f(x) = x^2 - 3x + 5$ on $[1, 4]$, if so find the value of c.

14. Let $f(x) = x + \frac{1}{x}$. To make a plot of $f'(x)$ and find the regions of increasing or decreasing and the local extrema of $f(x)$. Use a color-coded plot of $f(x)$ showing increasing or decreasing or local extrema.

15. Find the critical numbers, the regions of increasing or decreasing and the local extrema of $f(x) = -2x^3 + 6x^2 - 5$. Make a plot of $f(x)$ with the regions of increasing or decreasing color coded and the local extrema clearly labeled.

16. Using Maxima write Taylor series expansion of $x^3 + 7x^2 - 6$ in powers of $(x - 3)$.

17. Using Maxima write Taylor series expand e^x in powers of $(x - 2)$.

18. Using Maxima prove that $\tan^{-1} x = \frac{\pi}{4} + \frac{1}{2}(x - 1) - \frac{1}{4}(x - 1)^2 + \frac{1}{12}(x - 1)^3 + \cdots$.

19. Using Maxima prove that $\tan^{-1} x = x - \frac{x^3}{3} + \frac{x^5}{5} - \frac{x^7}{7} + \cdots$.

20. Using Maxima prove that $\tan^{-1}\left(\frac{1-x}{1+x}\right) = \frac{\pi}{4} - \frac{1}{2}\left(x - \frac{x^3}{3} + \frac{x^5}{5} - \frac{x^7}{7}\right) + \cdots$.

21. Using Maxima prove that $e^x \sin x = x + x^2 + \frac{2x^3}{3!} - \frac{4x^5}{5!} + \cdots$.

22. Using Maxima prove that $e^{\sin x} = 1 + x + \frac{1}{2}x^2 + \cdots$.

23. Using Maxima prove that $\log(1 + \sin x) = x - \frac{1}{2}x^2 + \frac{1}{3}x^3 - \frac{1}{12}x^4 + \cdots$.

———————————————◆◆◆———————————————

Chapter 3

Unit 3: Ordinary Differential Equations

3.1 Introduction

The most of students have knowledge of the calculus knowing all about differentiation and integration. Also, they have study some applications of derivative and integration. Many of the principles or laws underlying the behavior of the natural world are statements or relations involving rates at which things happen. When expressed in mathematical terms, the relations are equations and the rates are derivatives. In this chapter, we will extend these applications to those connected with differential equations. Differential equations are equations involving an unknown function and its derivatives. If the function is a function of a single variable, then the equations are known as ordinary differential equations. If the unknown function is a function of several independent variables, then the equation is a partial differential equation. In this chapter, we study the first order differential equations for which there are general methods of solution.

3.2 Linear First Order Equations

A first order differential equation is said to be *linear* if it can be written in standard form as

$$\frac{dy}{dx} + p(x)y = f(x). \tag{3.1}$$

An equation which cannot be written in this form is said to be *nonlinear*.

Now, (3.1) is said to be *homogeneous* if $f \equiv 0$; else it is *nonhomogeneous*.

Since $y \equiv 0$ is an obvious solution of the homogeneous equation

$$\frac{dy}{dx} + p(x)y = 0.$$

we call it the *trivial* solution. Any other solution is said to be *nontrivial*.

Remark 3.1. *Linear differential equations are those in which the dependent variable and its derivatives occur only in the first degree and are not multiplied together.*

Example 3.1. *Check whether the following first order equations are linear and if so, rewrite in*

(i) $x^2\frac{dy}{dx} + 3y = x^2$,   (ii) $x\frac{dy}{dx} - 8x^2y = sinx$

(iii) $x\frac{dy}{dx} + (lnx)y = 0$,   (iv) $\frac{dy}{dx} = x^2y - 2$.

Solution: The given first order equations are linear but not in form (3.1).

These can be rewritten in the standard form as

(i) $\frac{dy}{dx} + \frac{3}{x^2}y = 1$

(ii) $\frac{dy}{dx} - 8xy = \frac{sinx}{x}$

(iii) $\frac{dy}{dx} + \frac{(lnx)}{x}y = 0$

(iv) $\frac{dy}{dx} - x^2y = -2$.

Example 3.2. *The following are some nonlinear first order differential equations*

(i) $x\frac{dy}{dx} + 3y^2 = 2x$ *(because y is squared)*,

(ii) $y\frac{dy}{dx} = 3$ *(because of product yy')*,

(iii) $\frac{dy}{dx} + xe^y = 12$ *(because of e^y)*.

General Solution of a Linear First Order Equation

Consider the linear first order equation

$$\frac{dy}{dx} = \frac{1}{x^2}. \tag{3.2}$$

We can easily write this equation into variable separable form as follows

$$\frac{dy}{1} = \frac{dx}{x^2}$$

After integrating, we obtain the solution y of this equation as

$$y = \frac{-1}{x} + c. \tag{3.3}$$

where c is an arbitrary constant.

Remark 3.2. *We call c a parameter and say that* (3.3) *defines a* one-parameter family *of functions. For each real number c, the function defined by* (3.3) *is a solution of* (3.2) *on* $(-\infty, 0)$ *and* $(0, \infty)$. *Moreover, every solution of* (3.2) *on either of these intervals is of the form* (3.3) *for some choice of c. We say that* (3.3) *is the* general solution of (3.2).

We'll see that a similar situation occurs in connection with first order linear equation (3.1). *That is, if p and f are continuous on some open interval (a, b) then there's a unique formula $y = y(x, c)$ analogous to* (3.3) *which involves x and a parameter c and has the following properties:*

- *For each fixed value of c, the resulting function of x is a solution of* (3.1) *on* (a, b)

- *If y is a solution of (3.1) on (a, b), then y can be obtained from the formula by choosing c appropriately.*

We'll call $y = y(x, c)$ the general solution of (3.1).

When this has been established, it will follow that an equation of the form

$$P_0(x)\frac{dy}{dx} + P_1(x)y = F(x). \tag{3.4}$$

has a general solution on any open interval (a, b) on which P_0, P_1 and F are all continuous and P_0 has no zeros as we rewrite (3.4) in the form of (3.1) with $p = \frac{P_1}{P_0}$ and $f = \frac{F}{P_0}$, which are both continuous on (a, b). For completeness, we point out that if P_0, P_1 and F are all continuous on an open interval (a, b), but P_0 does have a zero in (a, b), then (3.4) may fail to have a general solution on (a, b) in the sense just defined and needs to be developed in depth.

3.2.1   Homogeneous Linear First Order Equations

Now, we obtain the general solution of a homogeneous linear first order equation.

Theorem 3.1. *If p is continuous on (a, b), then the general solution of the homogeneous equation*

$$\frac{dy}{dx} + p(x)y = 0 \tag{3.5}$$

on (a, b) is

$$y = ce^{-\phi(x)}$$

where

$$\phi(x) = \int p(x)dx \tag{3.6}$$

is any antiderivative of p on (a, b) that is,

$$\phi'(x) = p(x), \quad a < x < b. \tag{3.7}$$

Proof: Let $y = ce^{-\phi(x)}$,

Differentiating w.r.t. x, we get

$$\frac{dy}{dx} = -\phi'(x)ce^{-\phi(x)}$$

$$\frac{dy}{dx} = -p(x)ce^{-\phi(x)}. \quad \text{(using equation (3.7))}$$

$$= -p(x)y.$$

Therefore,

$$\frac{dy}{dx} + p(x)y = 0;$$

that is, y is a solution of (3.5), for any choice of c.

The trivial solution can be written as $y = ce^{-\phi(x)}$, with $c = 0$.

Now assume y is a nontrivial solution.

Then there is an open subinterval I of (a, b) on which y has no zeros.

We can rewrite (3.5) as

$$\frac{dy}{y} = -p(x)dx \tag{3.8}$$

for x in I.

Integrating (3.8) and recalling (3.6) yields

$$ln|y| = -\int p(x)dx + k,$$

where k is a constant.

$$ln|y| = -\phi(x) + k,$$

This implies that

$$|y| = e^k e^{-\phi(x)}.$$

Since P is defined for all x in (a, b) and an exponential can never be equal to zero, we can take $I = (a, b)$. Therefore, y has zeros on (a, b) and so we can rewrite that last equation as

$$y = ce^{-\phi(x)}$$

where

$$c = \begin{cases} e^k & \text{if } y > 0 \text{ on } (a, b) \\ -e^k & \text{if } y < 0 \text{ on } (a, b). \end{cases}$$

Example 3.3. *Let* a *be a constant.*

(a) Find the general solution of

$$\frac{dy}{dx} - ay = 0 \tag{3.9}$$

(b) Solve the initial value problem

$$\frac{dy}{dx} - ay = 0, \quad y(x_0) = y_0$$

Solution: (i) Consider the given homogeneous differential equation

$$\frac{dy}{dx} - ay = 0 \tag{3.10}$$

Comparing with

$$\frac{dy}{dx} + p(x)y = 0$$

$$\therefore p(x) = -a$$

Therefore, by theorem (3.1), the general solution of equation (3.9) is

$$y = ce^{-\phi(x)} \qquad (3.11)$$

where c is arbitrary constant and

$$\phi(x) = \int -a\,dx = -a \int dx$$
$$= -ax.$$

Therefore, equation (3.10) becomes

$$\therefore y = ce^{ax} \qquad (3.12)$$

where

$$c = \begin{cases} e^k & if\ y > 0\ on\ (a,b) \\ -e^k & if\ y < 0\ on\ (a,b). \end{cases}$$

This is the general solution of given equation.

This shows that every nontrivial solution of (3.9) is of the form $y = ce^{ax}$ for some nonzero

constant c. Now, setting $c = 0$ yields the trivial solution, *all* solutions of (3.9) are of the form (3.12).

Verification: Now, the equation (3.12) is a solution of (3.9) for every choice of c as differentiating

(3.12) yields

$$y' = ace^{ax} = ay.$$
$$\frac{dy}{dx} - ay = 0.$$

(b) Imposing the initial condition $y(x_0) = y_0$ gives

$$y_0 = ce^{ax_0}.$$

Therefore, $c = y_0 e^{-ax_0}$ and

$$y = y_0 e^{-ax_0} e^{ax} = y_0 e^{a(x-x_0)}.$$

In particular, the initial condition is $x_0 = 0$, $y_0 = 1$.

Putting in equation (3.12), we get

$$c = 1.$$

Therefore, the solution of given equation is

$$y = e^{ax} \qquad (3.13)$$

This is called a particular solution of the equation (3.9).

The graphs of the function $y = e^{ax}$ for various values of a are given below by Maxima software.

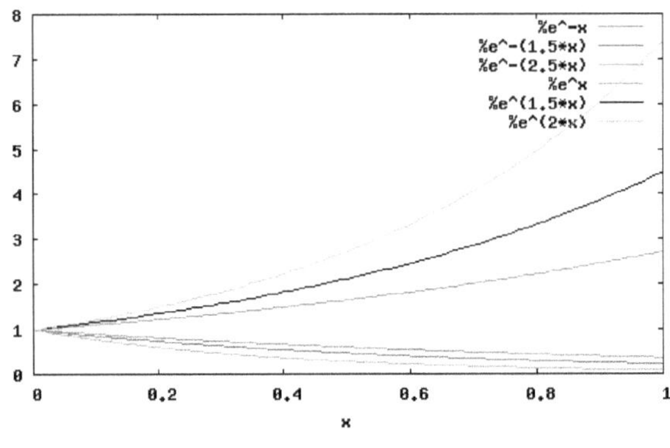

Figure 3.1

Example 3.4. *(i) Find the general solution of the homogeneous differential equation $xy' + y = 0$.*
(ii) Solve the initial value problem $xy' + y = 0$, $y(1) = 3$.

Solution: (i) Consider the given homogeneous differential equation

$$xy' + y = 0 \tag{3.14}$$

This can be written as

$$x\frac{dy}{dx} + y = 0$$

$$\frac{dy}{dx} + \frac{1}{x}y = 0$$

Comparing with

$$\frac{dy}{dx} + p(x)y = 0$$

$$\therefore p(x) = \frac{1}{x}.$$

Therefore, by theorem (3.80), the general solution of equation (4.49) is

$$y = ce^{-\phi(x)} \tag{3.15}$$

where c is arbitrary constant and

$$\phi(x) = \int p(x)dx = \int \frac{1}{x}dx$$

$$= \log x.$$

Therefore, equation (4.50) becomes

$$y = ce^{-\log x} = ce^{\log(1/x)}$$

$$\therefore y = \frac{c}{x} \tag{3.16}$$

where

$$\left\{ \begin{array}{ll} e^k & if \ y > 0 \ on \ (a,b) \end{array} \right.$$

This is the general solution of given equation.

(ii) We consider the initial condition $y(1) = 3$ that is $y_0 = 3$ when $x_0 = 1$.

Putting in equation (4.51), we get

$$c = 3$$

Therefore, $y = \frac{3}{x}$ is the solution of initial value problem.

This is called a particular solution of given homogeneous differential equation. The nature of the solation represented graphically by Maxima software as

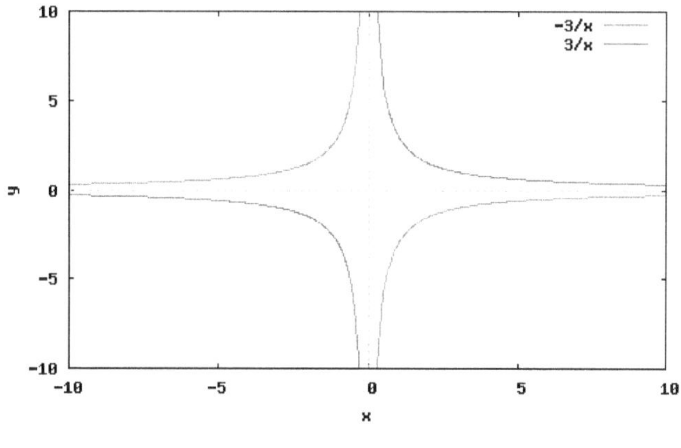

Figure 3.2

3.2.2 Linear Nonhomogeneous First Order Equations

In this section, we discuss the method of variation of parameters to obtain the solution of Linear Nonhomogeneous First Order Differential Equations.

Theorem 3.2. *The Method of Variation of Parameters (MVP)*

Suppose p and f are continuous on an open interval (a, b), and let y_1 be any nontrivial solution of the complementary equation

$$y' + p(x)y = 0 \quad on \ (a, b).$$

Then the general solution of the nonhomogeneous equation

$$y' + p(x)y = f(x), \quad on \ (a, b), \ is$$

$$y = y_1(x)\left\{ \int \frac{f(x)}{y_1(x)} dx + c \right\}.$$

Proof: We consider the linear nonhomogeneous differential equation

$$y' + p(x)y = f(x) \tag{3.17}$$

The complementary equation of equation (3.17) is

Suppose

$$y = uy_1 \tag{3.19}$$

is the solution of linear nonhomogeneous differential equation (3.17), where y_1 is a nontrivial solution of the complementary equation and u is to be determined.

Now, y_1 is a solution of complementary equation, therefore

$$y_1' + p(x)y_1 = 0 \tag{3.20}$$

This method of using a solution of the complementary equation to obtain solutions of a nonhomogeneous equation is a special case of a method called the method of variation of parameters.

If $y = uy_1$ then

$$y' = u'y_1 + uy_1' \tag{3.21}$$

Substituting these expressions for y and y' into (3.17) gives

$$u'y_1 + u(y_1' + p(x)y_1) = f(x),$$

which reduces to

$$u'y_1 = f(x) \tag{3.22}$$

We know that y_1 has no zeros on the interval. Therefore,

$$u' = \frac{f(x)}{y_1(x)} + c.$$

Now, to multiply the result by y_1 to get the general solution of (3.17).

Therefore,

$$y = y_1(x)\left(\int \frac{f(x)}{y_1(x)} dx + c \right)$$

where, y_1 is a solution of complementary equation.

Remark 3.3. *We summarize the method of variation of parameters for solving*

$$\frac{dy}{dx} + p(x)y = f(x) \tag{3.23}$$

as follows:

(a) *Find a function y_1 such that*

$$\frac{y_1'}{y_1} = -p(x).$$

For convenience, take the constant of integration to be zero.

(b) *Write*

$$y = uy_1 \tag{3.24}$$

(d) Integrate u' to obtain u, with an arbitrary constant of integration.

(e) Substitute u into (3.24) to obtain y.

Example 3.5. *Solve the initial value problem*

$$y' + 2y = x^3 e^{-2x}, \quad y(0) = 1. \tag{3.25}$$

Solution: The complementary equation of equation (3.25) is

$$y' + 2y = 0 \tag{3.26}$$

Comparing with

$$\frac{dy}{dx} + p(x)y = 0$$

$$\therefore p(x) = 2.$$

Therefore, by theorem (3.80), the general solution of the complementary equation is

$$y = ce^{-\phi(x)} \tag{3.27}$$

where c is arbitrary constant and

$$\phi(x) = \int p(x)dx = \int 2dx$$
$$= 2x.$$

Therefore, $y = ce^{-2x}$ is the general solution of the complementary equation.

Here, we consider $y_1 = e^{-2x}$ is solution of the complementary equation.

By the method of variation of parameters, we consider

$$y = uy_1 = ue^{-2x}$$

is the solution of equation (3.25), so that u is to be determined.

Differentiating w.r.t. x, we get

$$y' = u'e^{-2x} - 2ue^{-2x}.$$

$$\therefore y' + 2y = u'e^{-2x} - 2ue^{-2x} + 2ue^{-2x} = u'e^{-2x}$$

Putting in the equation (3.25), we get

$$u'e^{-2x} = x^3 e^{-2x}$$
$$u' = x^3$$
$$\therefore u = \frac{x^4}{4} + c$$

Putting in the equation $y = ue^{-2x}$, we get

$$y = e^{-2x}\left(\frac{x^4}{4} + c\right).$$

This is the general solution of (3.25).

On imposing initial condition $y(0) = 1$ gives $c = 1$.

Therefore the solution of initial value problem is $y = e^{-2x}\left(\frac{x^4}{4} + 1\right)$.

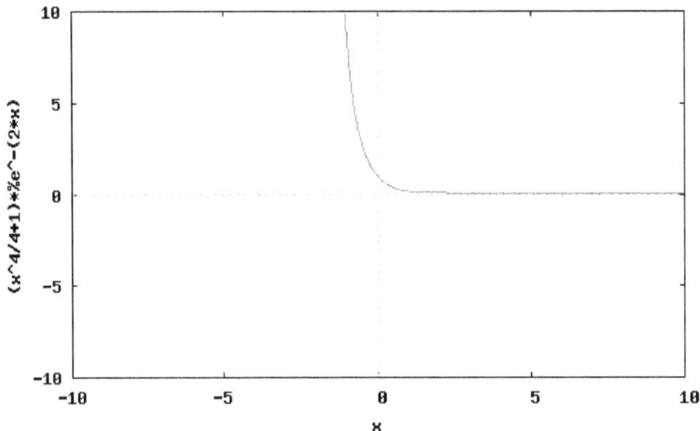

Figure 3.3: Integral curves

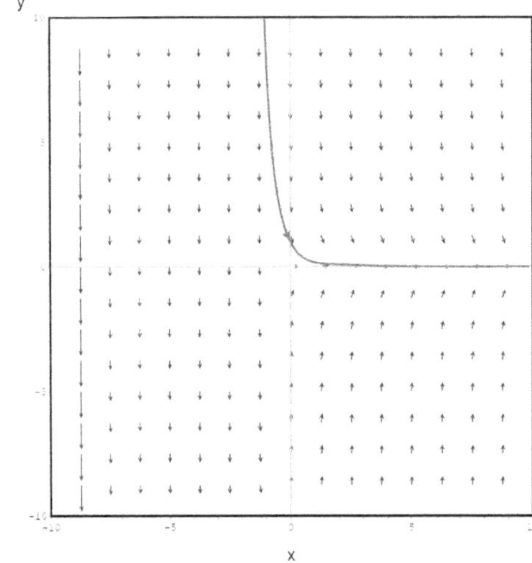

Figure 3.4: Direction fields

Remark 3.4. *Solutions in Integral Form*

Sometimes the integrals that arise in solving a linear first order equation can not be evaluated in terms of elementary functions. In this case the solution must be left in terms of an integral.

Example 3.6. *(a) Find the general solution of*

(b) Solve the initial value problem

$$y' - 2xy = 1, \quad y(0) = y_0. \tag{3.28}$$

Solution(a): To apply variation of parameters, we need a nontrivial solution y_1 of the complementary equation.

Thus, $y' - 2xy = 0$, is the complementary equation.

We choose $y_1 = e^{x^2}$ is solution of is the complementary equation.

Suppose $y = ue^{x^2}$ is the solution of (3.28).

$$y' - 2xy = u'e^{x^2} + 2xue^{x^2} - 2xy = u'e^{x^2} + 2xy - 2xy = u'e^{x^2} = 1$$

and hence,

$$u' = e^{-x^2}.$$

Therefore,

$$u = \int e^{-x^2}\, dx + c.$$

But we can not simplify the integral on the right because there is no elementary function with derivative equal to e^{-x^2}.

Therefore, the best available form for the general solution of (3.28) is

$$y = ue^{x^2} = e^{x^2}\left(\int e^{-x^2} dx + c \right). \tag{3.29}$$

(b) The initial condition in (3.28) is imposed at $x_0 = 0$, it is convenient to rewrite (3.29) as

$$y = e^{x_0^2}\left(\int_0^0 e^{-t^2} dt + c \right) = 0.$$

Setting $x = 0$ and $y = y_0$ here shows that $c = y_0$.

Therefore, the solution of the initial value problem is

$$y = e^{x^2}\left(\int_0^x e^{-x^2} dx + y_0 \right). \tag{3.30}$$

Existence and Uniqueness Theorem

The method of variation of parameters leads to this theorem.

Theorem 3.3. *Suppose p and f are continuous on an open interval (a, b), and let y_1 be any nontrivial solution of the complementary equation*

$$y' + p(x)y = 0$$

on (a, b). Then:

(a) The general solution of the nonhomogeneous equation

on (a, b) is

$$y = y_1(x)\left\{c + \int \frac{f(x)}{y_1(x)} dx\right\}.\tag{3.32}$$

(b) If x_0 is an arbitrary point in (a, b) and y_0 is an arbitrary real number, then the initial value problem

$$y' + p(x)y = f(x), \quad y(x_0) = y_0$$

has the unique solution

$$y = y_1(x)\left\{\frac{y_0}{y_1(x_0)} + \int_{x_0}^{x} \frac{f(t)}{y_1(t)} dt\right\}$$

on (a, b).

Proof (a) To show that (3.32) is the general solution of (3.31) on (a, b), we must prove that:

(i) If c is any constant, the function y in (3.32) is a solution of (3.31) on (a, b).

(ii) If y is a solution of (3.31) on (a, b) then y is of the form (3.32) for some constant c

To prove **(i)**, we first observe that any function of the form (3.32) is defined on (a, b), since p and f are continuous on (a, b). Differentiating (3.32) yields

$$y' = y_1'(x)\left\{c + \int \frac{f(x)}{y_1(x)} dx\right\} + f(x).$$

Since $y_1' = -p(x)y_1$, this and (3.32) imply that

$$y' = -p(x)y_1(x)\left\{c + \int \frac{f(x)}{y_1(x)} dx\right\} + f(x)$$
$$= -p(x)y(x) + f(x),$$

which implies that y is a solution of (3.31).

To prove **(ii)**, suppose y is a solution of (3.31) on (a, b). From the proof of Theorem (3.1), we know that y_1 has no zeros on (a, b), so the function $u = y/y_1$ is defined on (a, b). Moreover, since

$$y' = -py + f \text{ and } y_1' = -p\,y_1,$$

$$u' = \frac{y_1 y' - y_1' y}{y_1^2}$$
$$= \frac{y_1(-py + f) - (-py_1)y}{y_1^2}$$
$$= \frac{f}{y_1}$$

Integrating $u' = f/y_1$ gives

$$u = \left\{c + \int \frac{f(x)}{y_1(x)} dx\right\}$$

, which implies (3.32), since $y = uy_1$.

convenient to choose the antiderivative that equals zero when $x = x_0$, and write the general solution of (3.31) as

$$y = y_1(x)\left\{c + \int_{x_0}^x \frac{f(t)}{y_1(t)} dt\right\}.$$

Since

$$y(x_0) = y_1(x_0)\left\{c + \int_{x_0}^{x_0} \frac{f(t)}{y_1(t)} dt\right\} = cy_1(x_0),$$

we see that $y(x_0) = y_0$ if and only if $c = \frac{y_0}{y_1(x_0)}$.

Illustrative Examples

1. Find the general solution of the homogeneous differential equation $y' + 3x^2 y = 0$.

   Solution: The given homogeneous differential equation is

   $$y' + 3x^2 y = 0 \tag{3.33}$$

   Comparing with

   $$y' + p(x)y = 0$$

   $$\therefore p(x) = 3x^2$$

   Therefore, the general solution of equation is

   $$y = ce^{-\phi(x)} \tag{3.34}$$

   where c is arbitrary constant and

   $$\phi(x) = \int p(x)dx = \int 3x^2 dx$$
   $$= x^3.$$

   Therefore, putting in equation (3.34), we get

   $$y = ce^{-x^3}.$$

   This is the general solution of given equation.

2. Find the general solution of the homogeneous differential equation $xy' + (\log x)y = 0$.

   Solution: The given homogeneous differential equation is $xy' + (\log x)y = 0$.

   We write this as

   $$y' + \left(\frac{\log x}{x}\right)y = 0 \tag{3.35}$$

   Comparing with

   $$y' + p(x)y = 0$$
   $$\log x$$

Therefore, the general solution of equation is

$$y = ce^{-\phi(x)} \tag{3.36}$$

where c is arbitrary constant and

$$\phi(x) = \int p(x)dx = \int (\frac{\log x}{x}dx$$
$$= \int t\, dt \quad (\text{Put} \quad \log x = t)$$
$$= \frac{t^2}{2}$$
$$= \frac{(\log x)^2}{2}$$

Therefore, putting in equation (3.36), we get

$$y = ce^{-\frac{(\log x)^2}{2}}.$$

This is the general solution of given equation.

3. Solve the initial value problem $y' + (\frac{x+1}{x})y = 0$, $y(1) = 1$.

   Solution: Consider the given homogeneous differential equation

$$y' + (\frac{x+1}{x})y = 0 \tag{3.37}$$

Comparing with

$$y' + p(x)y = 0$$
$$\therefore p(x) = \frac{x+1}{x}$$

Therefore, the general solution of equation (3.37) is

$$y = ce^{-\phi(x)} \tag{3.38}$$

where c is arbitrary constant and

$$\phi(x) = \int p(x)dx = \int \frac{x+1}{x}dx$$
$$= \int (1 + \frac{1}{x})dx$$
$$= x + \log x.$$

Therefore, equation (3.38) becomes

$$y = ce^{-(x+\log x)}$$
$$\therefore y = \frac{ce^{-x}}{x} \tag{3.39}$$

where

$$c = \begin{cases} e^k & if\ y > 0\ on\ (a,b) \\ -e^k & if\ y < 0\ on\ (a,b). \end{cases}$$

This is the general solution of given equation.

We consider the initial condition $y(1) = 1$ that is $y_0 = 1$ when $x_0 = 1$.

Putting in equation (3.39), we get

$$c = e$$

Therefore, $y = \dfrac{e^{(x+1)}}{x}$ is the solution of initial value problem.

The nature of the solation represented graphically by Maxima software as

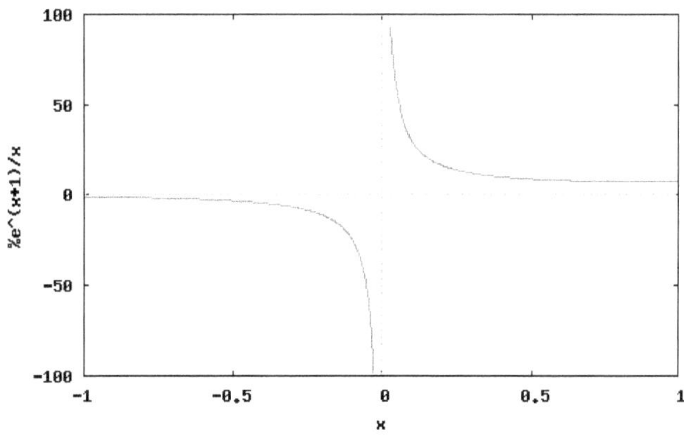

Figure 3.5

4. Solve the initial value problem $xy' + (1 + x\cot x)y = 0$, $y(\pi/2) = 2$.

   Solution: Consider the given homogeneous differential equation

$$xy' + (1 + x\cot x)y = 0 \tag{3.40}$$

We will write this as $y' + (\frac{1}{x} + \cot x)y = 0$

Comparing with $y' + p(x)y = 0$.

$$\therefore p(x) = \frac{1}{x} + \cot x$$

Therefore, the general solution of equation (3.40) is

$$y = ce^{-\phi(x)} \tag{3.41}$$

where c is arbitrary constant and

$$\phi(x) = \int p(x)dx = \int (\frac{1}{x} + \cot x)dx = \log x + \log \sin x = \log(x \sin x).$$

Therefore, equation (3.42) becomes

$$y = ce^{-\log(x \sin x)}$$

$$c$$

$$\tag{3.42}$$

This is the general solution of given equation.

We consider the initial condition $y(\pi/2) = 2$ that is $y_0 = 2$ when $x_0 = \pi/2$.

Putting in equation (3.42), gives $c = \pi$.

Therefore, $y = \frac{\pi}{x \sin x}$ is the solution of initial value problem.

The nature of the solution represented graphically by Maxima software as

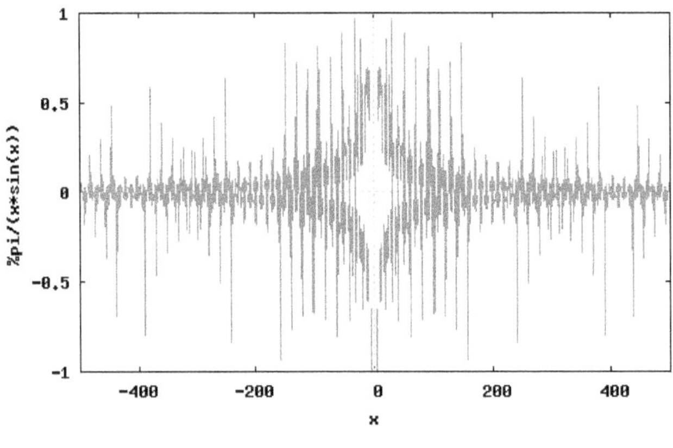

Figure 3.6

5. Find the general solution of following equation by method of variation of parameters

$$y' + (\frac{1}{x} - 1)y = \frac{-2}{x}.$$

Solution The given nonhomogeneous differential equation is

$$y' + (\frac{1}{x} - 1)y = \frac{-2}{x} \tag{3.43}$$

Therefore, comparing with $y' + p(x)y = f(x)$, we get $p(x) = \frac{1}{x} - 1$ and $f(x) = \frac{-2}{x}$ are both continuous except at the point $x = 0$.

We obtain the nontrivial solution y_1 of the complementary equation.

Thus, y_1 must satisfy the complementary equation $y' + (\frac{1}{x} - 1)y = 0$.

$$y' + (\frac{1}{x} - 1)y = 0 \tag{3.44}$$

Comparing with

$$y' + p(x)y = 0$$

$$\therefore p(x) = \frac{1}{x} - 1$$

Therefore,the general solution of the complementary equation is

$$y = ce^{-\phi(x)} \tag{3.45}$$

where c is arbitrary constant and

$$\phi(x) = \int p(x)dx = \int \left(\frac{1}{x} - 1\right) dx$$
$$= \frac{e^x}{x}.$$

Therefore, $y = c\frac{e^x}{x}$ is the general solution of the complementary equation.

Here, we consider $y_1 = \frac{e^x}{x}$ is solution of the complementary equation.

By the method of variation of parameters, we consider

$$y = uy_1 = \frac{ue^x}{x}$$

is the solution of given equation, so that u is to be determined.

Differentiating w.r.t. x, we get

$$y' = \frac{u'e^x}{x} + u\frac{xe^x - e^x}{x^2}.$$
$$\therefore y' + 2y = u'e^{-2x} - 2ue^{-2x} + 2ue^{-2x} = u'e^{-2x}$$

Putting in given nonhomogeneous equation, we get

$$u' = -2e^{-x}.$$

Integrating this gives

$$u = 2e^{-x} + c$$
$$\therefore y = \frac{e^x}{x}(2e^{-x} + c)$$

is the general solution of given nonhomogeneous equation.

6. Find the general solution of following equation by method of variation of parameters

$$y' + (\tan x)y = \cos x.$$

Solution The given nonhomogeneous differential equation is

$$y' + (\tan x)y = \cos x \tag{3.46}$$

Therefore, comparing with $y' + p(x)y = f(x)$, we get $p(x) = \tan x$ and $f(x) = \cos x$ are both continuous in the domain.

We obtain the nontrivial solution y_1 of the complementary equation.

Thus, y_1 must satisfy the complementary equation $y' + (\tan x)y = 0$.

$$y' + (\tan x)y = 0 \tag{3.47}$$

Comparing with

$$\therefore p(x) = \tan x$$

Therefore,the general solution of the complementary equation is

$$y = ce^{-\int p(x)dx} \tag{3.48}$$

where c is arbitrary constant and

$$y_1 = e^{-\int p(x)}dx = e^{\int \frac{-\sin x}{\cos x}} \, dx$$

$$= e^{\log(\cos x)} = \cos x.$$

Here, we consider $y_1 = \cos x$ is solution of the complementary equation.
By the method of variation of parameters, we consider

$$y = uy_1 = u \cos x$$

is the solution of given equation, so that u is to be determined.
Differentiating w.r.t. x, we get

$$y' = u' \cos x - u \sin x.$$

$$\therefore y' + (\tan x)y = u' \cos x - u \sin x + u \sin x = u' \cos x.$$

Putting in given nonhomogeneous equation, we get

$$u' = 1$$

Integrating this gives

$$u = x + c$$

$$\therefore y = (x + c) \cos x$$

is the general solution of given nonhomogeneous equation.

7. By the method of variation of parameters, solve the following initial value problem

$$y' + (\cot x)y = x cosec x, \ \ y\left(\frac{\pi}{2}\right) = 1.$$

Solution The given nonhomogeneous differential equation is

$$y' + (\cot x)y = x \, cosec \, x, \tag{3.49}$$

Therefore, comparing with $y' + p(x)y = f(x)$, we get $p(x) = \cot x$ and $f(x) = x \, cosec \, x$ are both
continuous except at the points $x = n\pi$, where n is an integer.

We obtain the nontrivial solution y_1 of the complementary equation.

Thus, y_1 must satisfy the complementary equation $y_1' + (\cot x)y_1 = 0$.

$$y' + (\cot x)y = 0 \tag{3.50}$$

Comparing with

$$y' + p(x)y = 0$$

$$\therefore p(x) = \cot x$$

Therefore,the general solution of the complementary equation is

$$y = ce^{-\phi(x)} \tag{3.51}$$

where c is arbitrary constant and

$$\phi(x) = \int p(x)dx = \int \cot x \ dx$$

$$= \log \sin x.$$

Therefore, $y = ce^{-\log \sin x} = \dfrac{c}{\sin x}$ is the general solution of the complementary equation.

Here, we consider $y_1 = \dfrac{1}{\sin x}$ is solution of the complementary equation.

By the method of variation of parameters, we consider

$$y = uy_1 = \frac{u}{\sin x}$$

is the solution of given equation, so that u is to be determined.

Differentiating w.r.t. x, we get

$$y' = \frac{u'}{\sin x} - \frac{u \cos x}{\sin^2 x}.$$

$$\therefore y' + 2y = u'e^{-2x} - 2ue^{-2x} + 2ue^{-2x} = u'e^{-2x}$$

Putting in given nonhomogeneous equation, we get

$$y' + (\cot x)y = \frac{u'}{\sin x} - \frac{u \cos x}{\sin^x} + \frac{u \cot x}{\sin x}$$

$$= \frac{u'}{\sin x} - \frac{u \cos x}{\sin^2 x} + \frac{u \cos x}{\sin^2 x}$$

$$= \frac{u'}{\sin x} \tag{3.52}$$

Therefore y is a solution of given nonhomogeneous equation if and only if

$$\frac{u'}{\sin x} = x \operatorname{cosec} x = \frac{x}{\sin x} \quad \text{or equivalently,} \quad u' = x.$$

Integrating this gives

$$x^2$$

$$\therefore y = \frac{u}{sin\,x} = \frac{x^2}{2sin\,x} + \frac{c}{sin\,x}. \tag{3.53}$$

is the general solution of given nonhomogeneous equation on every interval $(n\pi, (n+1)\pi)$.
Imposing the initial condition $y(\pi/2) = 1$ in (3.53) yields

$$1 = \frac{\pi^2}{8} + c \ or \ c = 1 - \frac{\pi^2}{8}$$

. Therefore,

$$y = \frac{x^2}{2sin\,x} + \frac{(1 - \frac{\pi^2}{8})}{sin\,x}$$

is a solution of given equation. The interval of validity of this solution is $(0, \pi)$.

8. solve the following initial value problem and sketch the graph of the solution

$$xy' - 2y = -x^2, \quad y(1) = 1.$$

Solution The given nonhomogeneous differential equation is

$$xy' - 2y = -x^2 \tag{3.54}$$

We write above equation as

$$y' - \frac{2}{x}y = -x$$

Therefore, comparing with $y' + p(x)y = f(x)$, we get $p(x) = -\frac{2}{x}$ and $f(x) = -x$ are both continuous in the domain.

We obtain the nontrivial solution y_1 of the complementary equation.

Thus, y_1 must satisfy the complementary equation $y' - \frac{2}{x}y = 0$.

$$y' - \frac{2}{x}y = 0 \tag{3.55}$$

Comparing with

$$y' + p(x)y = 0$$

$$\therefore p(x) = -\frac{2}{x}$$

Therefore,the general solution of the complementary equation is

$$y = ce^{-\int p(x)dx} \tag{3.56}$$

where c is arbitrary constant and

$$y_1 = e^{-\int p(x)dx} = e^{-\int -\frac{2}{x}dx}$$

$$= e^{\log(x^2)} = x^2.$$

Here, we consider $y_1 = x^2$ is solution of the complementary equation.

By the method of variation of parameters, we consider

$$y = uy_1 = ux^2$$

is the solution of given equation, so that u is to be determined.

Differentiating w.r.t. x, we get

$$y' = u'x^2 + 2ux.$$

$$\therefore xy' - 2y = u'x^3 + 2ux^2 - 2ux^2 = u'x^3.$$

Putting in given nonhomogeneous equation, we get

$$u' = -\frac{1}{x}$$

Integrating this gives

$$u = -\log x + c$$

$$\therefore y = x^2(c - \log x)$$

is the general solution of given nonhomogeneous equation. Imposing the initial condition $y(1) = 1$ that is $y = 1$ when $x = 1$ in above solution yields $c = 1$. Therefore,

$$y = x^2(1 - \log x)$$

is a solution of given equation.

Remark 3.5. *Note that some nonlinear equations can be transformed into linear equations by changing the dependent variable. If*

$$g'(y)y' + p(x)g(y) = f(x)$$

where y is a function of x and g is a function of y, then the new dependent variable $z = g(y)$ satisfies the linear equation $z' + p(x)z = f(x)$.

9. Solve the equation $(\sec^2 y)y' - 3\tan y = -1$.

   Solution The given nonlinear differential equation is

   $$(\sec^2 y)y' - 3\tan y = -1 \tag{3.57}$$

   Put $z = \tan y$, $z' = (\sec^y)y'$ Therefore, the given equation transformed into linear equation given by

   $$z' - 3z = -1 \tag{3.58}$$

   Therefore, comparing with $z' + p(x)z = f(x)$, we get $p(x) = -3$ and $f(x) = -1$ are both

We obtain the nontrivial solution z_1 of the complementary equation.

Thus, z_1 must satisfy the complementary equation $z' - 3z = 0$.

$$z' - 3z = 0 \tag{3.59}$$

Comparing with

$$z' + p(x)z = 0$$

$$\therefore p(x) = -3$$

Therefore, the general solution of the complementary equation is

$$y = ce^{-\int p(x)dx} \tag{3.60}$$

where c is arbitrary constant and

$$y_1 = e^{-\int p(x)dx} = e^{-\int 3dx}$$

$$= e^{3x}.$$

Here, we consider $y_1 = e^{3x}$ is solution of the complementary equation.

By the method of variation of parameters, we consider

$$z = uz_1 = ue^{3x}$$

is the solution of given equation, so that u is to be determined.

Differentiating w.r.t. x, we get

$$z' = u'e^{3x} + 3ue^{3x}.$$

$$\therefore z' - 3z = u'e^{3x} + 3ue^{3x} - 3ue^{3x} = u'e^{3x}.$$

Putting in given nonhomogeneous equation, we get

$$u' = -e^{-3x}$$

Integrating this gives

$$u = \frac{1}{3e^{3x}} + c$$

$$\therefore z = e^{3x}\left(\frac{1}{3e^{3x}} + c\right) = \frac{1}{3} + ce^{3x}$$

is the general solution of the nonhomogeneous equation in z and x. Putting $z = \tan y$ in above solution, we get

$$\tan y = \frac{1}{3} + ce^{3x}$$

Therefore, $y = \tan^{-1}(1/3 + ce^{3x})$ is the general solution of given equation.

1. Find the general solution of the following differential equations.

    i. $xy' + 3y = 0$,   Ans: $y = \frac{c}{x^3}$

    ii. $x^2 y' + y = 0$,   Ans: $y = c e^{\frac{1}{x}}$

2. solve the following initial value problem.

    i. $xy' + (1 + \frac{1}{\ln x})y = 0,\ y(e) = 1.$,   Ans: $y = \dfrac{e}{x \log(x)}$

    ii. $y' - \left(\dfrac{2x}{1+x^2}\right)y = 0,\ y(0) = 2.$ Ans: $y = 2x^2 + 2$

    iii. $y' + (\frac{k}{x})y = 0,\ y(1) = 3, (k\,constant).$ Ans: $y = 3e^{(-k\log(x))}$

    iv. $y' + (\tan k\,x)y = 0,\ y(0) = 2, (k\,constant).$ Ans : $y = 2 * e^{(\frac{-\log(\sec(kx))}{k})}$

3. Find the general solution of the following differential equations. Also, plot a direction field and
    some integral curves on the rectangular region $-2 \le x \le 2,\ \ -2 \le y \le 2$.

    i. $y' + 3y = 1.$ Ans: $y = e^{-3x}\left(\frac{e^{3x}}{3} + c\right)$

    ii. $y' + 2xy = xe^{-x^2}$ Ans: $y = (\frac{x^2}{2} + c)e^{-x^2}$

    iii. $y' + (\frac{2x}{1+x^2})y = \frac{e^{-x}}{1+x^2}$ Ans: $y = \frac{(c - e^{-x})}{(x^2+1)}$

4. Find the general solution of the following differential equations.

    i. $y' + \frac{1}{x}y = \frac{7}{x^2} + 3$ Ans: $y = \frac{7\log(x) + (3x^2 - 6x)/2 + 7/x + c)}{(x-1)}$

    ii. $xy' + (1 + 2x^2)y = x^3 e^{-x^2}$ Ans: $y = e^{-\frac{\log(x^2)}{2} - x^2}\left(\frac{x^3 |x|}{4} + c\right)$

    iii. $xy' + 2y = \frac{2}{x^2} + 1$

    iv. $(1 + x)y' + 2y = \frac{\sin x}{1+x}$ Ans: $y = \frac{c - \cos(x)}{(x+1)^2}$

    v. $(x - 1)(x - 2)y' - (4x - 3)y = (x - 2)^3$

    vi. $y' + (\sin 2x)y = e^{-\sin^2 x}$ Ans: $y = \left(\frac{x}{\sqrt{e}} + c\right) e^{\frac{\cos(2x)}{2}}$

    vii. $x^2 y' + 3xy = e^x$

5. Solve the following initial value problem and sketch the graph of the solution.

    i. $y' + 7y = e^{3x},\ y(0) = 0$ Ans: $y = e^{-7x}(e^{10x}/10 - 1)$

    ii. $(1 + x^2)y' + 4xy = \frac{2}{(1+x^2)},\ y(0) = 1$ Ans: $\frac{(2x+1)}{(x^4 + 2x^2 + 1)}$

    iii. $xy' + 3y = \frac{2}{(1+x^2)x},\ y(-1) = 0$ Ans: $y = \frac{\log(x^2+1) - \log(2)}{x^3}$

    iv. $y' + (\cot x)y = \cos x,\ y(\pi/2) = 1$ Ans: $y = -\frac{\cos(x)^2 - 2}{2\sin(x)}$

    v. $y' + \frac{1}{x}y = \frac{2}{x^2} + 1,\ y(-1) = 0$ Ans: $y = \frac{2\log(x^2) + x^2 - 1}{2x}$

vii. $y' + 2xy = x$, $y(0) = 3$ $\qquad$ Ans: $y = \dfrac{e^{-x^2}\left(e^{x^2}+5\right)}{2}$

6. Solve the initial value problem and leave the answer in a form involving a definite integral.

i. $y' + 2xy = x^2$, $y(0) = 3$ $\qquad$ Ans: $y = \dfrac{e^{-x^2}\left(\sqrt{\pi}\, i\,\mathrm{erf}(i\,x)+2\,x\,e^{x^2}+12\right)}{4}$

ii. $y' + \dfrac{2x}{1+x^2}y = \dfrac{e^x}{(1+x^2)^2}$, $y(0) = 1$

iii. $xy' + (x+1)y = e^{x^2}$, $y(1) = 2$. $\qquad$ Ans $y = -\dfrac{e^{-x-\frac{1}{4}}\left(\sqrt{\pi}\left(i\,\mathrm{erf}\left(\frac{2\,i\,x+i}{2}\right)-i\,\mathrm{erf}\left(\frac{3\,i}{2}\right)\right)-4\,e^{\frac{5}{4}}\right)}{2\,x}$

7. Solve by the following of the following differential equations.

i. $\dfrac{xy'}{y} + 2\ln y = 4x^2$ $\qquad$ Ans: $x^2 \log (y) - x^4 = c$

ii. $\dfrac{y'}{(1+y)^2} - \dfrac{1}{x(1+y)} = -\dfrac{3}{x^2}$ $\qquad$ Ans: $\dfrac{3\log(x)\,y+3\log(x)-x}{y+1} = c$

3.3   Separable Equations

A first order differential equation is *separable* if it can be written as

$$h(y)y' = g(x), \tag{3.61}$$

where the left side is product of y' and a function of y and the right side is a function of x. Expressing a separable differential equation in this form is called *separation of variables.* In section 3.1 we used separation of variables to solve homogeneous linear equations. In this section, we will apply this method to nonlinear equations.

To see how to solve (3.61), let's first assume that y is a solution. Let $G(x)$ and $H(y)$ be antiderivatives of $g(x)$ and $h(y)$; that is,

$$H'(y) = h(y) \ \ and \ \ G'(x) = g(x). \tag{3.62}$$

Then, from the chain rule,

$$\frac{d}{dx}H(y(x)) = H'(y(x))y'(x) = h(y)y'(x).$$

Therefore, (3.61) is equivalent to

$$\frac{d}{dx}H(y(x)) = \frac{d}{dx}G(x).$$

Integrating both sides of this equation and combining the constants of integration gives

$$H(y(x)) = G(x) + c. \tag{3.63}$$

Thus, any differentiable function y that satisfies (3.63) for some constant c is a solution of (3.61).

Example 3.7. *Solve the equation*

Solution: Separating variables yields

$$\frac{y'}{1 + y^2} = x.$$

Integrating gives

$$tan^{-1}y = \frac{x^2}{2} + c$$

. Therefore

$$y = tan(\frac{x^2}{2} + c).$$

Example 3.8. *(a) Solve the equation*

$$y' = -\frac{x}{y}. \tag{3.64}$$

(b) Solve the initial value problem

$$y' = -\frac{x}{y}, \quad y(1) = 1. \tag{3.65}$$

(c) Solve the initial value problem

$$y' = -\frac{x}{y}, \quad y(1) = -2. \tag{3.66}$$

Solution:(a) Separating variables in (3.64) gives

$$ydy = -xdx.$$

Integration gives

$$\frac{y^2}{2} = -\frac{x^2}{2} + c, \ i.e., \ x^2 + y^2 = 2c.$$

This shows that c must be positive if y is to be a solution of (3.64) on an open interval. Therefore we let $2c = a^2$ with $(a > 0)$ and rewrite the last equation as

$$x^2 + y^2 = a^2. \tag{3.67}$$

This equation has two differentiable solutions for y in terms of x :

$$y = \sqrt{a^2 - x^2}, \quad -a < x < a, \tag{3.68}$$

and

$$y = -\sqrt{a^2 - x^2}, \quad -a < x < a, \tag{3.69}$$

The solution curves defined by (3.68) are semicircles above the $x-axis$ and those defined by (3.69) are semicircles below the $x-axis$.

Solution:(b) The solution of (3.65) is positive when $x = 1$; hence, it is of the form (3.68). Substituting $x = 1$ and $y = 1$ into (3.67) to satisfy the initial condition yields $a^2 = 2$; hence, the solution of (3.65) is

$$y = \sqrt{2 - x^2}, \quad -\sqrt{2} < x < \sqrt{2}.$$

Solution:(c) The solution of (3.66) is negative when $x = 1$ and is therefore of the form (3.69). Substituting $x = 1$ and $y = -2$ into (3.67) to satisfy the initial condition gives $a^2 = 5$. Hence, the solution of (3.66) is

$$y = -\sqrt{5 - x^2}, \quad -\sqrt{5} < x < \sqrt{5}.$$

3.3.1   Implicit Solutions of Separable Equations

In above examples, we solved the equation $H(y) = G(x) + c$ to obtain explicit formulas for solutions of the given separable differential equations. This is not possible always. Therefore, we enrich our definition of a solution of a separable equation in the next theorem.

Theorem 3.4. *Suppose $g = g(x)$ is continuous on (a, b) and $h = h(y)$ is continuous on (c, d). Let G be an antiderivative of g on (a, b) and let H be an antiderivative of h on (c, d). Let x_0 be an arbitrary point in (a, b), let y_0 be a point in (c, d) such that $h(y_0) \neq 0$, and define*

$$c = H(y_0) - G(x_0). \tag{3.70}$$

Then there's a function $y = y(x)$ defined on some open interval $(a_1, b_1,)$ where $a \leq a_1 < x_0 < b_1 \leq b$, such that $y(x_0) = y_0$ and

$$H(y) = G(x) + c \tag{3.71}$$

for $a_1 < x < b_1$. Therefore y is a solution of the initial value problem

$$h(y)y' = g(x), \quad y(x_0) = x_0. \tag{3.72}$$

It is convenient to say that (3.71) with c arbitrary is an *implicit solution* of $h(y)y' = g(x)$. Curves defined by (3.71) are integral curves of $h(y)y' = g(x)$.

If c satisfies (3.70), we will say that (3.71) is an *implicit solution of the initial value problem* (3.72).

(i) For some choices of c there may not be any differentiable functions y that satisfy (3.71).

(ii) The function y in (3.71) is a solution of $h(y)y' = g(x)$.

Example 3.9. *(i) Find implicit solutions of $y' = \frac{2x+1}{5y^4+1}, \quad y(2) = 1.$*

Solution(a) The given differential equation is

$$y' = \frac{2x + 1}{5y^4 + 1} \tag{3.73}$$

Now, separating variables, we get

$$(5y^4 + 1)y' = 2x + 1.$$

Integrating gives the implicit solution of (3.73) as

Imposing the initial condition $y(2) = 1$ in (3.74) gives $c = -4$.

Therefore,

$$y^5 + y = x^2 + x - 4$$

is an implicit solution of the initial value problem. Now, more than one differentiable function $y = y(x)$ satisfies (3.73) near $x = 1$, it can be shown that there is only one such function that satisfies the initial condition $y(1) = 2$.

3.3.2 Constant Solutions of Separable Equations

An equation of the form

$$y' = g(x)p(y)$$

is separable, since it can be rewritten as

$$\frac{1}{p(y)}y' = g(x).$$

However, the division by $p(y)$ is not legitimate if $p(y) = 0$ for some values of y. The next two examples show how to tackle with this problem.

Example 3.10. *Find all solutions of*

$$y' = 2xy^2 \tag{3.75}$$

Solution Let $y' = 2xy^2$ be a given differential equation.

Dividing by y^2 that separate the variables, therefore, we have

$$\frac{y'}{y^2} = 2x.$$

That is of the form

$$\frac{1}{p(y)}y' = 2x$$

Therefore, $p(y) = y^2$ to separate variables.

We observe that $y = 0$ is one of the solution of given equation but it creates difficulty in case of separate variables.

Now, suppose y is a solution of (3.75) that is not identically zero.

Since y is continuous there must be an interval on which y is never zero.

Therefore, division by y^2 is allowed for x in this interval, so we can separate variables and to obtain

$$\frac{y'}{y^2} = 2x.$$

Integrating this gives

$$-\frac{1}{y} = x^2 + c,$$

which is equivalent to

$$y = -\frac{1}{x^2 + c}.$$ (3.76)

We have shown that if y is a solution of the given that is not identically zero, then y must be of the form

$$y = -\frac{1}{(x^2 + c)}.$$

Example 3.11. *Find all solutions of*

$$\frac{dy}{dx} = \frac{1}{2}x(1 - y^2).$$ (3.77)

Solution Let $y' = \frac{1}{2}x(1 - y^2)$ be a given differential equation.

Dividing by y^2 that separate the Variables, therefore, we have

$$\frac{y'}{1 - y^2} = \frac{1}{2}x.$$

That is of the form

$$\frac{1}{p(y)}y' = \frac{1}{2}x$$

Therefore, $p(y) = 1 - y^2$ to separate variables.

We observe that $y = \pm 1$ is one of the solution of given equation but it creates difficulty in case of separate variables.

Therefore, this solution will obtain only by inspection. Now, suppose y is a solution of the given equation that is not identically zero.

Since y is continuous there must be an interval on which y is never zero.

Therefore, division by $1 - y^2$ is allowed for x in this interval, so we can separate variables and to obtain

$$\frac{y'}{1 - y^2} = \frac{1}{2}x.$$

Now suppose y is a solution of given equation such that $1 - y^2$ is not identically zero.

Therefore, $1 - y^2$ is continuous there must be an interval on which $1 - y^2$ is never zero.

We can write above equation as

$$\left[\frac{1}{y - 1} - \frac{1}{y + 1}\right] dy = -x dx.$$

Integrating, we get

$$ln\left|\frac{y - 1}{y + 1}\right| = -\frac{x^2}{2} + k$$

$$\therefore \left|\frac{y - 1}{y + 1}\right| = e^k e^{-\frac{x^2}{2}}.$$

Therefore, we can rewrite this equation as

$$\frac{y - 1}{\qquad} \qquad {}^2/2$$

where $c = \pm e^k$, depending upon the sign of $(y-1)/(y+1)$ on the interval.

Now,solving for y gives

$$y = \frac{1 + ce^{-x^2/2}}{1 - ce^{-x^2/2}}.$$ (3.78)

Difference Between Linear and Nonlinear Equations

We know that the theorem (3.2) states that if $p(x)$ and $f(x)$ are continuous on (a,b) then every solution of

$$y' + p(x)y = f(x)$$

on (a,b) can be obtained by choosing a value for the constant c in the general solution, and if x_0 is any point in (a,b) and y_0 is arbitrary, then the initial value problem

$$y' + p(x)y = f(x), \quad y(x_0) = y_0$$

has a solution on (a,b).

This is not true for nonlinear equations. A nonlinear equation may have solutions that cant be obtained by choosing a specific value of a constant appearing in a one-parameter family of solutions. It is in general impossible to estimate the interval of validity of a solution to an initial value problem for a nonlinear equation by simply examining the equation, as the interval of validity may depend on the initial condition.

Example 3.12. *Solve the initial value problem*

$$y' = 2xy^2, \quad y(0) = y_0$$

and determine the interval of validity of the solution.

Solution Now, suppose that $y_0 \neq 0$. Therefore, the given equation can be written as

$$\frac{1}{y^2}dy = 2xdx$$

. Integrating, we get

$$-\frac{1}{y} = x^2 + c$$

. i.e., the solution y is in the form of

$$y = -\frac{1}{x^2 + c}$$ (3.79)

Now, imposing the initial condition gives $c = -\frac{1}{y_0}$.

Substituting this into (3.79) and rearranging terms gives the solution

$$y = \frac{y_0}{1 - y_0 x^2}.$$

This is also the solution if $y_0 = 0$.

If $y_0 > 0$, then $1 - y_0 x^2 = 0$ gives $x = \pm \dfrac{1}{\sqrt{y_0}}$.

Hence, the solution is valid only on $\left(-\dfrac{1}{\sqrt{y_0}}, \dfrac{1}{\sqrt{y_0}} \right)$.

Illustrative Examples

Example 3.13. *Find all solutions of the following nonlinear equations.*

   1. $y' = \dfrac{3x^2 + 2x + 1}{y - 2}$

   2. $(sin\,x)(sin\,y) + (cos\,y)y' = 0$

Solution (i) Let $y' = \dfrac{3x^2 + 2x + 1}{y - 2}$ be a given nonlinear differential equation. On separation of variables, we get

$$(y - 2)dy = (3x^2 + 2x + 1)dx$$

Integrating, we get

$$\frac{y^2}{2} - 2y = x^3 + x^2 + x + c$$

This is the general solution of given equation.

(ii) Let $(sin\,x)(sin\,y) + (cos\,y)y' = 0$ be a given nonlinear differential equation. On separation of variables, we get

$$sin\,x\,dx + \frac{cos\,y}{sin\,y}dy = 0$$

Integrating, we get

$$-\cos x + \log \sin y = .c$$

This is the general solution of given equation.

Example 3.14. *Find all solutions of the following differential equation. Also, plot a direction field and some integral curves on the indicated rectangular region.*

$$y' = x^2(1 + y^2), \quad -1 \le x \le 1, \ -1 \le y \le 1.$$

Solution Let $y' = x^2(1 + y^2))$ be a given differential equation.

Dividing by $1 + y^2$ that separate the Variables, therefore, we have

$$\frac{y'}{1 + y^2} = x^2.$$

That is of the form

$$\frac{1}{p(y)}y' = x^2$$

Therefore, $p(y) = 1 + y^2$ to separate variables.

Since y is continuous there must be an interval on which y is never zero.

Therefore, division by $1 + y^2$ is allowed for x in this interval, so we can separate variables and obtain

$$\frac{y'}{1 + y^2} = x^2.$$

Now, suppose y is a solution of given equation such that $1 + y^2$ is not identically zero.

Therefore, $1 + y^2$ is continuous there must be an interval on which $1 + y^2$ is never zero.

We can write above equation as

$$\frac{dy}{1 + y^2} = x^2 dx$$

Integrating, we get

$$\tan^{-1} y = \frac{x^3}{3} + c$$

$$\therefore y = \tan\left(\frac{x^3}{3} + c\right)$$

is the general solution.

Example 3.15. *solve the following initial value problem and find the interval of validity of the solution*

$$y' = -2x(y^2 - 3y + 2), \ y(0) = 3.$$

Solution Let $y' = -2x(y^2 - 3y + 2)$ be a given differential equation.

Dividing by $y^2 - 3y + 2$ that separate the Variables, therefore, we have

$$\frac{y'}{y^2 - 3y + 2} = -2x.$$

That is of the form

$$\frac{1}{p(y)}y' = \frac{1}{2}x$$

Therefore, $p(y) = y^2 - 3y + 2 = (y - 1)(y - 2)$ to separate variables.

We observe that $y = 1, 2$ are solutions of given equation but it creates difficulty in case of separate variables.

Therefore, this solution will obtain only by inspection. Now, suppose y is a solution of the given equation that is not identically zero.

Since y is continuous there must be an interval on which y is never zero.

Therefore, division by $y^2 - 3y + 2$ is allowed for x in this interval, so we can separate variables and to obtain

$$\frac{y'}{y^2 - 3y + 2} = -2x.$$

Now suppose y is a solution of given equation such that $y^2 - 3y + 2$ is not identically zero.

Therefore, $y^2 - 3y + 2$ is continuous there must be an interval on which $y^2 - 3y + 2$ is never zero.

We can write above equation as

$$y'$$

$$\left[\frac{1}{y-2}-\frac{1}{y-1}\right]dy = -2xdx.$$

Integrating, we get

$$ln\left|\frac{y-2}{y-1}\right| = -x^2 + c$$

$$\therefore \left|\frac{y-2}{y-1}\right| = e^k e^{(-x^2+c)}.$$

Therefore, we can rewrite this equation as

$$\frac{y-2}{y-1} = ce^{(-x^2+c)},$$

where $c = \pm e^k$, depending upon the sign of $(y-2)/(y-1)$ on the interval.

Now,solving for y gives

$$y = \frac{2 - e^{(-x^2+c)}}{1 - e^{(-x^2+c)}}.$$

Using initial condition $y(0) = 3$, we obtain $c = \log(0.5)$

Putting in above equation, we get

$$y = \frac{2 - e^{(-x^2+\log(0.5))}}{1 - e^{(-x^2+\log(0.5))}}.$$

The solution is valid on $(-\infty, \infty)$.

Example 3.16. *Solve the following equation using variation of parameters followed by separation of variables assuming that $p(x) \neq 0$*

$$y' + y = \frac{2xe^{-x}}{1+ye^x}.$$

Solution(a): The given differential equation is

$$y' + y = \frac{2xe^{-x}}{1+ye^x}$$

$p(x) = 1$ and $f(x) = \dfrac{2xe^{-x}}{1+ye^x}$ are both continuous in the domain.

We obtain the nontrivial solution y_1 of the complementary equation.

Thus, y_1 must satisfy the complementary equation $y' + y = 0$.

$$y' + y = 0$$

Comparing with

$$y' + p(x)y = 0$$

$$\therefore p(x) = 1$$

Therefore,the general solution of the complementary equation is

$$y = ce^{-\int p(x)dx}$$

where c is arbitrary constant and

$$y_1 = e^{-\int p(x)dx} = e^{-\int dx}$$

$$= e^{-x}.$$

Here, we consider $y_1 = e^{-x}$ is solution of the complementary equation.
By the method of variation of parameters, we consider

$$y = uy_1 = ue^{-x}$$

is the solution of given equation, so that u is to be determined.
Differentiating w.r.t. x, we get

$$y' = u'e^{-x} - ue^{-x}.$$

$$\therefore y' + y = u'e^{-x} - ue^{-x} + ue^{-x} = u'e^{-x}.$$

Putting in given nonhomogeneous equation, we get

$$u' = -\frac{2x}{1+u}$$

By separating the variables, we get

$$(1+u)du = 2xdx$$

Integrating this gives

$$u + \frac{u^2}{2} = x^2 + c$$

Substituting $u = ye^x$, we ontain

$$ye^x + \frac{y^2e^x}{2} = x^2 + c.$$

This the general solution of given equation.

Exercise 3.2

1. Find all solutions of the following equations.

    i. $xy' + y^2 + y = 0$ Ans: $\log(y+1) - \log(y) = \log(x) + c$

    ii. $y'ln|y| + x^2y = 0$ Ans: $-\frac{\log(|y|)^2}{2} = \frac{x^3}{3} + c$

    iii. $(3y^3 + 3y\cos y + 1)y' + \frac{2x+1}{1+x^2}y = 0$, Ans: $-3\sin(y) - \log(y) - y^3 = \log(x^2+1) + \text{atan}(x) + c$

    iv. $x^2yy' = (y^2 - 1)^{3/2}$ Ans: $-\frac{1}{\sqrt{y^2-1}} = c - \frac{1}{x}$

2. Find all solutions of the following equations. Also, plot a direction field and some integral curves on the indicated rectangular region.

ii. $y' = (x-1)(y-1)(y-2);\ [-2 \le x \le 2, -3 \le y \le 3]$

iii. $(y-1)^2 y' = 2x + 3;\ [-2 \le x \le 2, -2 \le y \le 5]$

3. Solve the initial value problem.

i. $y' = \dfrac{x^2 + 3x + 2}{y - 2},\ \ y(1) = 4$
   Ans: $\dfrac{y^2 - 4y}{2} = \dfrac{2x^3 + 9x^2 + 12x - 23}{6}$

ii. $y' + x(y^2 + y) = 0,\ \ y(2) = 1$
    Ans: $\log(y+1) - \log(y) = \dfrac{x^2 + 2\log(2) - 4}{2}$

4. Solve the initial value problem and graph the solution.

i. $y' + \dfrac{(y+1)(y-1)(y-2)}{(x+1)} = 0,\ \ y(1) = 0$

ii. $y' + 2x(y+1) = 0,\ \ y(0) = 2$

iii. $y' = 2xy(y^2 + 1),\ \ y(0) = 1$
     Ans: $-\dfrac{\log(y^2 + 1) - \log(y^2)}{4} = \dfrac{2x^2 - \log(2)}{4}$

5. Solve the initial value problem and find the interval of validity of the solution.

i. $y' = -2x(y^2 - 3y + 2),\ \ y(0) = 3$
   Ans: $\dfrac{\log(y-1) - \log(y-2)}{2} = \dfrac{x^2 + \log(2)}{2}$

ii. $y' = \dfrac{2x}{1+2y},\ \ y(2) = 0$
    Ans: $\dfrac{y^2 + y}{2} = \dfrac{x^2 - 4}{2}$

iii. $y' = 2y - y^2,\ \ y(0) = 1$
     Ans: $\dfrac{\log(y) - \log(y-2)}{2} = \dfrac{2x - \log(-1)}{2}$

iv. $x + yy' = 0,\ \ y(3) = -4$
    Ans: $-\dfrac{y^2}{2} = \dfrac{x^2 - 25}{2}$

v. $y' + x^2(y+1)(y-2)^2 = 0,\ \ y(4) = 1,$ Ans: $-\dfrac{(y-2)\log(y+1) + (2-y)\log(y-2) - 3}{9y - 18} = \dfrac{3x^3 - \log(2) + \log(-1) - 195}{9}$

6. $(x+1)(x-2)y' + y = 0,\ \ y(1) = -3$

6. Solve the following equations using variation of parameters followed by separation of variables (assuming that $p(x) \ne 0$).

i. $xy' - 2y = \dfrac{x^6}{y + x^2}$
   Ans: $\dfrac{y^2 + 2x^2 y - x^6}{2x^4} = c$

ii. $y' - y = \dfrac{(x+1)e^{4x}}{(y + e^x)^2}$
    Ans: $\dfrac{e^{-3x}\left(y^3 + 3e^x y^2 + 3e^{2x} y - 3xe^{4x}\right)}{3} = c$

iii. $y' - 2y = \dfrac{xe^{2x}}{1 - ye^{-2x}}$
     Ans: $\dfrac{e^{-4x}\left(y^2 - 2e^{2x} y + x^2 e^{4x}\right)}{2} = -\dfrac{1}{2}.$

3.4 Existence and Uniqueness of Solution of Nonlinear Equations

There are some methods for solving few nonlinear equations but for most most of nonlinear equations it is impossible to deduce useful formulas to obtain those solutions. Therefore, it becomes important and useful to know the conditions that imply the existence and uniqueness of solutions (exact or numerical solutions) of initial value problems for nonlinear equations. The next theorem gives sufficient conditions for existence and uniqueness of solutions of initial value problems for first order nonlinear

Theorem 3.5. *(a) If f is continuous on an open rectangle*

$$R : (a < x < b,\ c < y < d)$$

that contains (x_0, y_0) then the initial value problem

$$y' = f(x, y),\ \ y(x_0) = y_0 \tag{3.80}$$

has at least one solution on some open subinterval of (a, b) that contains x_0.

(b) If both f and f_y are continuous on R then (3.80) has a unique solution on some open subinterval of (a, b) that contains x_0.

Example 3.17. *Consider the initial value problem*

$$y' = \frac{x^2 - y^2}{x^2 + y^2},\ \ y(x_0) = y_0. \tag{3.81}$$

Here

$$f(x, y) = \frac{x^2 - y^2}{x^2 + y^2}\ \text{ and }\ f_y(x, y) = -\frac{4x^2 y}{(x^2 + y^2)^2}$$

are continuous everywhere except at $(0, 0)$.

If $(x_0, y_0) \neq (0, 0)$, there is an open rectangle R that contains (x_0, y_0) that does not contain $(0, 0)$. Since f and f_y are continuous on R, Theorem (3.80) implies that if $(x_0, y_0) \neq (0, 0)$ then (3.81) has a unique solution on some open interval that contains x_0.

Example 3.18. *Consider the initial value problem*

$$y' = \frac{10}{3} x y^{\frac{2}{5}},\ \ y(x_0) = y_0 \tag{3.82}$$

(a) For what points (x_0, y_0) does Theorem (3.80) (a) imply that (3.82) has a solution ?

(b) For what points (x_0, y_0) does Theorem (3.80) (b) imply that (3.82) has a unique solution on some open interval that contains x_0 ?

Solution:(a) Since

$$f(x, y) = \frac{10}{3} x y^{\frac{2}{5}}$$

is continuous for all (x, y), Theorem (3.80) implies that (3.82) has a solution for every (x_0, y_0).

Solution:(b) Here

$$f_y(x, y) = \frac{4}{3} x y^{-\frac{3}{5}}$$

is continuous for all (x, y) with $y \neq 0$.

Therefore, if $y_0 \neq 0$ there's an open rectangle on which both f and f_y are continuous, and Theorem (3.80) implies that (3.82) has a unique solution on some open interval that contains x_0.

If $y = 0$ then $f_y(x, y)$ is undefined, and therefore discontinuous; hence, Theorem (3.80) does not

Example 3.19. *Consider the initial value problem*

$$y' = \frac{x^2 - y^2}{1 + x^2 + y^2}, \quad y(x_0) = y_0. \tag{3.83}$$

Since

$$f(x, y) = \frac{x^2 - y^2}{1 + x^2 + y^2} \quad and \quad f_y(x, y) = -\frac{2y(1 + 2x^2)}{(1 + x^2 + y^2)^2}$$

are continuous for all (x, y), Theorem (3.80) implies that if (x_0, y_0) is arbitrary, then (3.83) has a unique solution on some open interval that contains x_0.

Example 3.20. *Consider the initial value problem*

$$y' = \frac{x + y}{x - y}, \quad y(x_0) = y_0. \tag{3.84}$$

Here

$$f(x, y) = \frac{x + y}{x - y} \quad and \quad f_y(x, y) = \frac{2x}{(x - y)^2}$$

are continuous everywhere except on the line $y = x$.

If $y_0 \neq x_0$, there's an open rectangle R that contains (x_0, y_0) that does not intersect the line $y = x$. Since f and f_y are continuous on R, Theorem (3.80) implies that if $y_0 \neq x_0$, (3.84) has a unique solution on some open interval that contains x_0.

Exercise 3.3

1. Find all (x_0, y_0) for which theorem (3.5) implies that initial value problem $y' = f(x, y)$, $y(x_0) = y_0$ has **(1)** a solution **(2)** a unique solution on some open interval that contains x_0.

   (i) $y' = \dfrac{x^2 + y^2}{\sin x}$,

   (ii) $y' = \dfrac{e^x + y}{x^2 + y^2}$,

   (iii) $y' = \tan x \, y$,

   (iv) $y' = \dfrac{x^2 + y^2}{\ln x \, y}$,

   (v) $y' = (x^2 + y^2) \, y^{1/3}$,

   (vi) $y' = 2 \, x \, y$,

   (vii) $y' = \ln(1 + x^2 + y^2)$.

$\bullet\bullet\bullet$

Chapter 4

Unit 4: Exact Differential Equations

4.1 Transformation of Nonlinear Equations into Separable Equations

In previous chapter, we found that the solution of a linear nonhomogeneous equation

$$y' + p(x)y = f(x) \tag{4.1}$$

is of the form $y = uy_1$, where y_1 is a nontrivial solution of the complementary equation

$$y' + p(x)y = 0 \tag{4.2}$$

and u is a solution of

$$u'y_1(x) = f(x).$$

Note that this equation is separable, since it can be rewritten as

$$u' = \frac{f(x)}{y_1(x)}.$$

In this section we will consider nonlinear differential equations that are separable to begin with, but can be solved in a similar way by writing their solutions in the form $y = uy_1$, where y_1 is a suitably chosen known function and u satisfies a separable equation. We will say in this case that we transformed the given equation into a separable equation.

4.1.1 Bernoulli Equations

A Bernoulli equation is an equation of the form

$$y' + p(x)y = f(x)y^n \tag{4.3}$$

where n can be any real number other than 0 or 1.

We can transform (4.3) into a separable equation by variation of parameters.

If y_1 is a nontrivial solution of the complementary equation

$$y' + p(x)y = 0$$

$$y_1' + p(x)y_1 = 0$$

Then by substituting $y = uy_1$ into (4.3) gives

$$u'y_1 + uy_1' + p(x)uy_1 = f(x)(uy_1)^n$$

$$u'y_1 + u(y_1' + p(x)y_1) = f(x)(uy_1)^n,$$

which is equivalent to the separable equation

$$u'y_1(x) = f(x)(y_1(x))^n u^n$$

$$\frac{u'}{u^n} = f(x)(y_1(x))^{n-1}.$$

Integrating above equation, we get

$$\int \frac{du}{u^n} = \int f(x)(y_1(x))^{n-1} dx$$

This gives the solution in function $u(x)$ and variable x.

Now, substitute $u = \frac{y}{y_1}$ in above solution, we will obtain the solution y for Bernoulli Equation (4.3).

Example 4.1. *Solve the Bernoulli equation*

$$y' - y = xy^2. \tag{4.4}$$

Solution(a): The given differential equation is

$$y' - y = xy^2$$

Let y_1 be the solution of complementary equation $y' - y = 0$.

$$y_1 = e^{-\int p(x)dx} = e^{\int dx}$$

$$= e^x.$$

We have to obtain the solutions of equation (4.4) using MVP.

Let $y = ue^x$ be solution of above equation.

$$y' = u'e^x + ue^x$$

$$y' - y = u'e^x + ue^x - ue^x = xu^2 e^{2x}$$

$$u'e^x = xu^2 e^{2x}$$

$$\therefore u' = xu^2 e^x.$$

Now, separating variables yields

$$\frac{u'}{u^2} = xe^x.$$

On integrating, we obtain

Hence,

$$u = -\frac{1}{(x-1)e^x + c}$$

Resubstituting, $u = \frac{y}{e^x}$, we obtain Therefore,

$$y = -\frac{1}{x - 1 + ce^{-x}}$$

is the general solution of the Bernoulli equation.

Other nonlinear equations that can be transformed into Separable equations

We have seen that the nonlinear Bernoulli equation can be transformed into a separable equation by the substitution $y = uy_1$ if y_1 is suitably chosen.

We now discover a sufficient condition for a nonlinear first order differential equation

$$y' = f(x,y) \tag{4.5}$$

to be transformable into a separable equation in the same way.

Substituting $y = uy_1$ into (4.5) yields

$$u'y_1(x) + uy_1'(x) = f(x, uy_1(x)),$$

which is equivalent to

$$u'y_1(x) = f(x, uy_1(x)) - uy_1'(x). \tag{4.6}$$

If

$$f(x, uy_1(x)) = q(u)y_1'(x)$$

for some function q, then (4.6) becomes

$$u'y_1(x) = (q(u) - u)y_1'(x), \tag{4.7}$$

which is separable.

After checking for constant solutions $u \equiv u_0$ such that $q(u_0) = u_0$, we can separate variables to obtain

$$\frac{u'}{q(u) - u} = \frac{y_1'(x)}{y_1(x)}.$$

Integrating, we get

$$\int \frac{du}{q(u) - u} = \int \frac{y_1'(x)}{y_1(x)} dx + c$$

$$\int \frac{du}{q(u) - u} = \log[y_1(x)] + c$$

is the solution of the first order nonlinear equation.

4.1.2   Homogeneous Nonlinear Equations

The nonlinear first order differential equation

$$y' = f(x, y) \tag{4.8}$$

is said to be *homogeneous* if x and y occur in f in such a way that $f(x, y)$ depends only on the ratio $\frac{y}{x}$; that is, (4.8) can be written as

$$y' = q\left(\frac{y}{x}\right), \tag{4.9}$$

where $q = q(u)$ is a function of a single variable.

Example 4.2. *Solve*

$$y' = \frac{y + xe^{-y/x}}{x}. \tag{4.10}$$

Solution The given differential equation is

$$y' = \frac{y + xe^{-y/x}}{x}.$$

This can be write as

$$y' = \frac{y}{x} + e^{-\frac{y}{x}} \tag{4.11}$$

That is of the form

$$y' = q\left(\frac{y}{x}\right)$$

Therefore the given equation is homogeneous.

We put $y = ux$, therefore, $q(u) = u + e^{-u}$ and $y' = u'x + u$.

Therefore, equation (4.11) becomes

$$y' = u + e^{-u}$$

$$\therefore u'x + u = u + e^{-u}$$

$$u'x = e^{-u}$$

$$e^u u' = \frac{1}{x}$$

This separate the variables and integrating, we get

$$\int e^u du = \int \frac{1}{x} dx + c$$

$$\therefore e^u = \log x + c$$

$$u = \log(\log x + c)$$

$$\frac{y}{x} = \log(\log x + c)$$

$$\therefore y = x \log(\log x + c)$$

Example 4.3. *Solve the initial value problem*

$$x^2 y' = y^2 + xy - x^2, \ y(1) = 2.$$

Solution The given differential equation is

$$x^2 y' = y^2 + xy - x^2.$$

This can be write as

$$y' = \left(\frac{y}{x}\right)^2 + \frac{y}{x} - 1 \tag{4.12}$$

That is of the form

$$y' = q\left(\frac{y}{x}\right)$$

Therefore the given equation is homogeneous.

We put $y = ux$, therefore, $q(u) = u + e^{-u}$ and $y' = u'x + u$.

Therefore, equation (4.12) becomes

$$y' = u^2 + u - 1$$

$$\therefore u'x + u = u^2 + u - 1$$

$$u'x = u^2 - 1$$

$$\frac{u'}{u^2 - 1} = \frac{1}{x}$$

$$\frac{du}{u^2 - 1} = \frac{1}{x}dx$$

This separate the variables and integrating, we get

$$\int \frac{du}{u^2 - 1}du = \int \frac{1}{x}dx + \log c$$

$$\frac{1}{2}\left[\frac{1}{u - 1} - \frac{1}{u + 1}\right] = \log x + \log c$$

$$\frac{1}{2}\log\left(\frac{u - 1}{u + 1}\right) = \log cx)$$

$$\frac{u - 1}{u + 1} = cx^2$$

$$u = \frac{1 + cx^2}{1 - cx^2}$$

$$\therefore y = \frac{x(1 + cx^2)}{(1 - cx^2)}$$

is the general solution of given homogeneous equation. Now, imposing the initial condition $y(1) = 2$ on above solution, we obtain $c = 0$.

Therefore, particular solution of given equation is $y = x$.

<u>**Illustrative Examples**</u>

Example 4.4. *Solve the Bernoulli equation*

$$y' + y = y^2. \tag{4.13}$$

Solution: The given differential equation is

$$y' + y = y^2$$

Let y_1 be the solution of complementary equation $y' + y = 0$.

$$y_1 = e^{-\int p(x)dx} = e^{-\int dx}$$

$$= e^{-x}.$$

We have to obtain the solutions of equation (4.13) using MVP.

Let $y = ue^{-x}$ be solution of above equation.

$$y' = u'e^{-x} - ue^{-x}$$

$$y' + y = u'e^{-x} - ue^{-x} + ue^{-x} = u^2e^{-2x}$$

$$u'e^{-x} = u^2e^{-2x}$$

$$\therefore u' = u^2e^{-x}.$$

Now, separating variables yields

$$\frac{u'}{u^2} = e^{-x}.$$

On integrating, we obtain

$$-\frac{1}{u} = -e^{-x} + c.$$

Hence,

$$u = -\frac{1}{e^{-x} + c}$$

Resubstituting, $u = ye^x$, we obtain Therefore,

$$y = \frac{1}{1 + ce^x}$$

is the general solution of the Bernoulli equation.

Example 4.5. *Solve the initial value problem*

$$y' - 2y = xy^3, \quad y(0) = 2\sqrt{2}. \tag{4.14}$$

Solution: The given differential equation is

Let y_1 be the solution of complementary equation $y' - 2y = 0$.

$$y_1 = e^{-\int p(x)dx} = e^{\int 2dx}$$

$$= e^{2x}.$$

We have to obtain the solutions of equation (4.14) using the method of variation of parameters.

Let $y = ue^{2x}$ be solution of above equation.

$$y' = u'e^{2x} + 2ue^{2x}$$

$$y' - 2y = u'e^{2x} + 2ue^{2x} - 2ue^{2x} = xu^3e^{6x}$$

$$u'e^{2x} = xu^3e^{6x}$$

$$\therefore u' = xu^3e^{4x}.$$

Now, separating variables yields

$$\frac{u'}{u^3} = xe^{4x}.$$

On integrating, we obtain

$$-\frac{1}{2u^2} = \frac{xe^{4x}}{4} - \frac{e^{4x}}{16} + c.$$

$$-\frac{x^2}{2y^2} = \frac{xe^{4x}}{4} - \frac{e^{4x}}{16} + c.$$

Imposing initial condition $y(0) = 2\sqrt{2}$ gives $c = \frac{1}{8}$.

Therefore, $y = \frac{x^2}{y^2} = \frac{e^{4x}}{2}\left(x - \frac{1}{4}\right) + \frac{1}{8}$ is the particular solution of the given Bernoulli equation.

Example 4.6. *Solve*

$$y' = \frac{y}{x} + sec(\frac{y}{x}). \tag{4.15}$$

Solution: The given differential equation is

$$y' = \frac{y}{x} + sec(\frac{y}{x}).$$

That is of the form

$$y' = q(\frac{y}{x})$$

Therefore the given equation is homogeneous.

We put $y = ux$, therefore, $q(u) = u + secu$ and $y' = u'x + u$.

Therefore, equation (4.15) becomes

$$y' = u + secu$$

$$\therefore u'x + u = u + secu$$

$$u'x = secu$$

$$\frac{u'}{secu} = \frac{1}{x}$$

$$udu$$

This separate the variables and integrating, we get

$$\int \cos u\, du = \int \frac{1}{x} dx + c$$
$$\sin u = \log x + c$$
$$\therefore \sin\left(\frac{y}{x}\right) = \log x + c$$

is the general solution of given homogeneous equation.

Exercise 4.1

1. Solve the following Bernoulli equations.

   i. $7xy' - 2y = -\frac{x^2}{y^6}$,   Ans: $\frac{y^7 - x}{x^2} = c$

   ii. $x^2 y' + 2y = 2e^{1/x} y^{1/2}$ Ans: $y = \left(c - \frac{1}{x}\right)^2 e^{\frac{2}{x}}$

   iii. $(1 + x^2)y' + 2xy = \frac{1}{(1+x^2)y}$ Ans: $\frac{\left(x^4 + 2x^2 + 1\right)y^2 - 2x}{2} = c$

2. Find all solutions. Also, plot a direction field and some integral curves on the indicated rectangular region.

   i. $y' - xy = x^3 y^3$; $-3 \le x \le 3, -2 \le y \le 2$

   ii. $y' - y\frac{1+x}{3x} = y^4$; $-2 \le x \le 2, -2 \le y \le 2$

3. Solve the initial value problem.

   i. $y' - xy = xy^{3/2}$, $y(1) = 4$ Ans: $\log(y) - 2\log\left(\sqrt{y} + 1\right) = \frac{x^2 + 2\log(4) - 4\log(3) - 1}{2}$

   ii. $xy' + y = x^4 y^4$, $y(1) = 1/2$ Ans: $y = \frac{1}{(11 - 3x)^{\frac{1}{3}} x}$

   iii. $y' - 2y = 2y^{1/2}$, $y(0) = 1$ Ans: $\log\left(\sqrt{y} + 1\right) = x + \log(2)$

   iv. $y' - 4y = \frac{48x}{y^2}$, $y(0) = 1$

   v. $y' = \frac{xy + y^2}{x^2}$, $y(-1) = 2$ Ans: $-\sqrt{e}\,x = e^{-\frac{x}{y}}$

   vi. $y' = \frac{x^3 + y^3}{xy^2}$, $y(1) = 3$ Ans: $\frac{y^3 - x^3 \log\left(x^3\right)}{3x^3} = 9$

   vii. $xyy' + x^2 + y^2 = 0$, $y(1) = 2$

   viii. $y' = \frac{y^2 - 3xy - 5x^2}{x^2}$, $y(1) = -1$

   ix. $x^2 y' = 2x^2 + y^2 + 4xy$, $y(1) = 1$

   x. $xyy' = 3x^2 + 4y^2$, $y(1) = \sqrt{3}$

4. Solve the initial value problem and graph the solution.

   i. $\quad {}^2\,{}' + 2 \quad\quad {}^3 \quad (1) = 1/\sqrt{2}$ Ans: $\quad \sqrt{5}$

ii. $y' - y = xy^{1/2}, \; y(0) = 4$ Ans: $y = x^2 + 4x + 4$

5. Solve the equation explicitly.

   i. $y' = \dfrac{y + x}{x}$

   ii. $y' = \dfrac{y^2 + 2xy}{x^2}$

   iii. $xy^3 y' = y^4 + x^4$

   iv. $y' = \frac{y}{x} + sec\left(\frac{y}{x}\right)$

6. Solve the equation explicitly. Also, plot a direction field and some integral curves on the indicated rectangular region.

   i. $x^2 y' = xy + x^2 + y^2; \; -8 \le x \le 8, -8 \le y \le 8$

   ii. $xyy' = x^2 + 2y^2; \; -4 \le x \le 4, -4 \le y \le 4$

   iii. $y' = \dfrac{2y^2 + x^2 e^{-(y/x)^2}}{2xy}; \; -8 \le x \le 8, -8 \le y \le 8$

7. Solve the given homogeneous equation implicitly.

   i. $y' = \dfrac{y + x}{x - y}$

   ii. $(xy' - y)(\ln|y| - \ln|x|) = x$

   iii. $y' = \dfrac{y^3 + 2xy^2 + x^2 y + x^3}{x(x + y)^2}$

   iv. $y' = \dfrac{x + 2y}{2x + y}$

   v. $y' = \dfrac{y}{y - 2x}$

   vi. $y' = \dfrac{xy^2 + 2y^3}{x^3 + x^2 y + xy^2}$

   vii. $y' = \dfrac{x^3 + x^2 y + 3y^3}{x^3 + 3xy^2}$

8. Find a function y_1 such that the substitution $y = uy_1$ transforms the given equation into a separable equation and solve the given equation explicitly.

   i. $3xy^2 y' = y^3 + x$ Ans: $\dfrac{y^3 - x\log(x)}{x} = c$

   ii. $xyy' = 3x^6 + 6y^2$ Ans: $\dfrac{y^2 + x^6}{2x^{12}} = c$

   iii. $x^3 y' = 2(y^2 + x^2 y - x^4)$ Ans: $x = ce^{-\frac{\log\left(\frac{y+x^2}{x^2}\right) - \log\left(\frac{y-x^2}{x^2}\right)}{4}}$

4.2   Exact Equations

The first order differential equation is of the form

$$M(x, y)dx + N(x, y)dy = 0 \tag{4.16}$$

This equation can be interpreted as

$$M(x, y) + N(x, y)\frac{dy}{dx} = 0 \tag{4.17}$$

where x is the independent variable and y is the dependent variable, or as

$$M(x, y)\frac{dx}{dy} + N(x, y) = 0 \tag{4.18}$$

where y is the independent variable and x is the dependent variable.

Since the solutions of equations of (4.17) and (4.18) will often have to be left in implicit form, we will say that $F(x, y) = c$ is an implicit solution of (4.16) if every differentiable function $y = y(x)$ that satisfies $F(x, y) = c$ is a solution of (4.17) and every differentiable function $x = x(y)$ that satisfies $F(x, y) = c$ is a solution of (4.18).

Here are some examples:

Example 4.7. *Show that*

$$x^4 y^3 + x^2 y^5 + 2xy = c \tag{4.19}$$

is an implicit solution of

$$(4x^3 y^3 + 2xy^5 + 2y)dx + (3x^4 y^2 + 5x^2 y^4 + 2x)dy = 0 \tag{4.20}$$

Solution Considering y as a function of x and differentiating (4.19) implicitly with respect to x yields

$$(4x^3 y^3 + 2xy^5 + 2y) + (3x^4 y^2 + 5x^2 y^4 + 2x)\frac{dy}{dx} = 0.$$

Similarly, regarding x as a function of y and differentiating (4.19) implicitly with respect to y yields

$$(4x^3 y^3 + 2xy^5 + 2y)\frac{dx}{dy} + (3x^4 y^2 + 5x^2 y^4 + 2x) = 0.$$

Therefore, (4.19) is an implicit solution of (4.20) in either of its two possible interpretations.

However, it illustrates the next important theorem, which we will prove by using implicit differentiation.

Partial Derivatives

Let $z = f(x, y)$ be a function of two variables x and y, where x and y are independent variables and z is the dependent variable.

We use the symbol ∂ instead of d and introduce the partial derivatives of z (or f) which are respect to x holding y constant.

(ii) $\frac{\partial z}{\partial y}$ or $\frac{\partial f}{\partial y}$ is read as partial derivative of z (or f) with respect to y that means differentiate with respect to y holding x constant.

(iii) Another common notations for partial derivatives:

(a) z_x means $\frac{\partial z}{\partial x}$ or $\frac{\partial f}{\partial x}$,

(b) z_y means $\frac{\partial z}{\partial y}$ or $\frac{\partial f}{\partial y}$. (iv) Second order partial derivatives:

Let $z = f(x, y)$ be a function of two variables x and y. (a) $\frac{\partial^2 z}{\partial x^2}$ i.e. z_{xx} or $\frac{\partial^2 f}{\partial x^2}$ (f_{xx}) means second derivative with respect to x holding y constant.

(b) $\frac{\partial^2 z}{\partial y^2}$ i.e. z_{yy} or $\frac{\partial^2 f}{\partial y^2}$ (f_{yy}) means second derivative with respect to y holding x.

(c) $\frac{\partial^2 z}{\partial x \partial y}$ means differentiate first with respect to y then with respect to x. these are called mixed partial derivatives.

The following are some examples.

Example 4.8. *Calculate* $\frac{\partial z}{\partial x}, \frac{\partial z}{\partial x}, \frac{\partial^2 z}{\partial x^2}, \frac{\partial^2 z}{\partial y^2}$ *and* $\frac{\partial^2 z}{\partial x \partial y}$ *when* $z = x^2 + 3xy + 2y - 7$.

Solution: To find $\frac{\partial z}{\partial x}$ treat y as a constant and differentiate w.r.t.x.

We have

$$z = x^2 + 3xy + 2y - 7$$
$$\frac{\partial z}{\partial x} = 2x + 3y$$
$$\frac{\partial z}{\partial y} = 3x + 2$$

$$\frac{\partial^2 z}{\partial x^2} = \frac{\partial}{\partial x}\left(\frac{\partial z}{\partial x}\right) = \frac{\partial}{\partial x}(2x + 3y) = 2$$
$$\frac{\partial^2 z}{\partial y^2} = \frac{\partial}{\partial y}\left(\frac{\partial z}{\partial y}\right) = \frac{\partial}{\partial y}(3x + 2) = 0$$
$$\frac{\partial^2 z}{\partial x \partial y} = \frac{\partial}{\partial x}\left(\frac{\partial z}{\partial y}\right) = \frac{\partial}{\partial x}(3x + 2) = 3$$
$$\frac{\partial^2 z}{\partial y \partial x} = \frac{\partial}{\partial y}\left(\frac{\partial z}{\partial x}\right) = \frac{\partial}{\partial y}(2x + 3y) = 3.$$

Example 4.9. *Calculate partial derivatives of the following functions.*

(i) $z = 4x^2 - 8xy^4 + 7y^5 - 9$,   *(ii)* $z = x\cos(xy)$.

Theorem 4.1. *If* $F = F(x, y)$ *has continuous partial derivatives* F_x *and* F_y, *then*

$$F(x, y) = c, \quad (c \text{ is constant}) \tag{4.21}$$

is an implicit solution of the differential equation

Proof: Regarding y as a function of x and differentiating (4.21) implicitly with respect to x yields

$$F_x(x, y) + F_y(x, y)\frac{dy}{dx} = 0.$$

On the other hand, regarding x as a function of y and differentiating (4.21) implicitly with respect to y yields

$$F_x(x, y)\frac{dx}{dy} + F_y(x, y) = 0.$$

Thus, (4.21) is an implicit solution of (4.22) in either of its two possible interpretations.

We will say that the equation

$$M(x, y)dx + N(x, y)dy = 0 \tag{4.23}$$

is *exact* on an open rectangle $\mathbb{R}$ if there's a function $F = F(x, y)$ such F_x and F_y, are continuous, and

$$F_x(x, y) = M(x, y) \ \ and \ \ F_y(x, y) = N(x, y) \tag{4.24}$$

for all (x, y) in $\mathbb{R}$.

This usage of "exact" is related to its usage in calculus, where the expression

$$F_x(x, y)dx + F_y(x, y)dy$$

is the exact differential of F.

Now, example (4.16) shows that it is easy to solve (4.23) if it is exact and we know a function F that satisfies (4.24).

The connected query would arise as:

Given an equation (4.23), how can we determine whether it is exact and if exact, how do we find a function F satisfying (4.24) ?

To analyse the answer, assume that there is a function F that satisfies (4.24) on some open rectangle $\mathbb{R}$, and in addition that F has continuous mixed partial derivatives F_{xy} and F_{yx}.

Then a theorem from calculus implies that

$$F_{xy} = F_{yx}. \tag{4.25}$$

If $F_x = M$ and $F_y = N$, differentiating the first of these equations with respect to y and the second with respect to x yields

$$F_{xy} = M_y \ \ and \ \ F_{yx} = N_x \tag{4.26}$$

From (4.25) and (4.26), we conclude that a necessary condition for exactness is that $M_y = N_x$.

This motivates the next theorem, which is stated as below.

Theorem 4.2. *The Exactness Condition*

Suppose M and N are continuous and have continuous partial derivatives M_y and N_x on an open rectangle $\mathbb{R}$. Then *is exact on R if and only if*

$$M_y(x, y) = N_x(x, y) \tag{4.27}$$

for all (x, y) in $\mathbb{R}$.

Example 4.10. *Show that the equation $3x^2y\,dx + 4x^3\,dy = 0$ is not exact on any open rectangle.*

Solution The given equation is $3x^2y\,dx + 4x^3\,dy = 0$.

Comparing with $M(x, y)dx + N(x, y)dy = 0$, we get

$M(x, y) = 3x^2y$ and $N(x, y) = 4x^3$.

So, $M_y = 3x^2$ and $N_x = 12x^2$.

Therefore, $M_y = N_x$ on the line $x = 0$, but not on any open rectangle, so there is no function F such that $F_x(x, y) = M(x, y)$ and $F_y(x, y) = N(x, y)$ for all (x, y) on any open rectangle.

The next example illustrates two possible methods for finding a function F that satisfies the condition $F_x = M$ and $F_y = N$ if $M\,dx + N\,dy = 0$ is exact.

Example 4.11. *Solve*

$$(4x^3y^3 + 3x^2)dx + (3x^4y^2 + 6y^2)dy = 0 \tag{4.28}$$

Solution: Method I: The given equation is $(4x^3y^3 + 3x^2)dx + (3x^4y^2 + 6y^2)dy = 0$.

Comparing with $M(x, y)dx + N(x, y)dy = 0$, we get

$$M(x, y) = 4x^3y^3 + 3x^2, \quad N(x, y) = 3x^4y^2 + 6y^2,$$

and

$$M_y(x, y) = N_x(x, y) = 12x^3y^2$$

for all (x, y).

The given equation is exact. Therefore, theorem (4.2) implies that there is a function F such that

$$F_x(x, y) = M(x, y) = 4x^3y^3 + 3x^2 \tag{4.29}$$

and

$$F_y(x, y) = N(x, y) = 3x^4y^2 + 6y^2 \tag{4.30}$$

for all (x, y).

To find F, we integrate (4.29) with respect to x to obtain

$$F(x, y) = x^4y^3 + y^3 + \phi(y), \tag{4.31}$$

where $\phi(y)$ is the "constant" of integration. If ϕ is any differentiable function of y then F satisfies equation (4.29).

with respect to y.

This yields

$$F_y(x, y) = 3x^4 y^2 + \phi'(y).$$

Comparing this with (4.30) shows that

$$\phi'(y) = 6y^2.$$

We integrate this with respect to y and take the constant of integration to be zero because we are interested only in finding some F that satisfies (4.29) and (4.30).

This yields

$$\phi(y) = 2y^3.$$

Substituting this into (4.31) yields

$$F(x, y) = x^4 y^3 + x^3 + 2y^3. \tag{4.32}$$

Now, theorem (4.1) implies that

$$x^4 y^3 + x^3 + 2y^3 = c$$

is an implicit solution of (4.28).

Solving this for y yields the explicit solution

$$y = \left(\frac{c - x^3}{2 + x^4} \right)^{\frac{1}{3}}.$$

Method II: Instead of first integrating (4.29) with respect to x, we could begin by integrating (4.30) with respect to y to obtain

$$F(x, y) = x^4 y^3 + 2y^3 + \psi(x). \tag{4.33}$$

where ψ is an arbitrary function of x. To determine ψ, we assume that ψ is differentiable and differentiate F with respect to x, which yields

$$F_x(x, y) = 4x^3 y^3 + \psi'(x).$$

Comparing this with (4.29) shows that

$$\psi'(y) = 3x^2.$$

Integrating this and again taking the constant of integration to be zero yields

$$\psi(x) = x^3.$$

Substituting this into equation (4.33) yields

$$F(\quad) = \quad^{4}\;^{3} + \;^{3} + 2\;^{3}$$

Now, theorem (4.1) implies that

$$x^4 y^3 + x^3 + 2y^3 = c$$

is an implicit solution of (4.28).

Solving this for y yields the explicit solution

$$y = \left(\frac{c - x^3}{2 + x^4} \right)^{\frac{1}{3}}.$$

Procedure For Solving An Exact Equation

Step 1. Check that the equation

$$M(x, y)dx + N(x, y)dy = 0 \tag{4.34}$$

satisfies the exactness condition $M_y = N_x$. If not then do not proceed further with this procedure.

Step 2. Integrate

$$\frac{\partial F(x, y)}{\partial x} = M(x, y)$$

with respect to x to obtain

$$F(x, y) = G(x, y) + \phi(y), \tag{4.35}$$

where G is an antiderivative of M with respect to X, and ϕ is an unknown function of y.

Step 3. Differentiate (4.35) with respect to y to obtain

$$\frac{\partial F(x, y)}{\partial y} = \frac{\partial G(x, y)}{\partial y} + \phi'(y)$$

Step 4. Equate the right side of this equation to N and solve for ϕ'; thus,

$$\frac{\partial G(x, y)}{\partial y} + \phi'(y) = N(x, y), \;\; so \;\; \phi'(y) = N(x, y) - \frac{\partial G(x, y)}{\partial y}.$$

Step 5. Integrate ϕ' with respect to y, taking the constant of integration to be zero, and substitute the result in (4.35) to obtain $F(x, y)$.

Step 6. Set $F(x, y) = c$ to obtain an implicit solution of (4.34). If possible, solve for y explicitly as a function of x.

Remark 4.1. *Many equations can be conveniently solved by either of the two methods used in just previous example. However, sometimes the integration required in one approach is more difficult than in the other. In such cases we choose the approach that requires the easier integration.*

Example 4.12. *Solve the equation*

$$(ye^{xy}\tan x + e^{xy}\sec^2 x)dx + xe^{xy}\tan x\, dy = 0. \tag{4.36}$$

Solution The given equation is $(ye^{xy}\tan x + e^{xy}\sec^2 x)dx + xe^{xy}\tan x dy = 0$.

Comparing with $M(x,y)dx + N(x,y)dy = 0$, we get

$$M(x,y) = ye^{xy}\tan x + e^{xy}\sec^2 x, \quad N(x,y) = xe^{xy}\tan x,$$

$$M_y(x,y) = e^{xy}\tan x + xye^{xy}\tan x + xe^{xy}\sec^2 x$$

$$N_x(x,y) = e^{xy}\tan x + xye^{xy}\tan x + xe^{xy}\sec^2 x$$

$$\therefore M_y(x,y) = N_x(x,y), \ \forall \ (x,y).$$

Therefore, the given equation is exact. We must find a function F such that

$$F_x(x,y) = ye^{xy}tanx + e^{xy}\sec^2 x \tag{4.37}$$

and

$$F_y(x,y) = xe^{xy}tanx \tag{4.38}$$

It is difficult to integrate (4.37) with respect to x, but easy to integrate (4.38) with respect to y. This gives

$$F(x,y) = e^{xy}tanx + \psi(x). \tag{4.39}$$

Differentiating this with respect to x yields

$$F_x(x,y) = ye^{xy}tanx + e^{xy}\sec^2 x + \psi'(x).$$

Comparing this with (4.37) shows that $\psi'(x) = 0$. Hence, ψ is a constant, which we can take to be zero in (4.39), and

$$e^{xy}tanx = c$$

is an implicit solution of (4.36).

Example 4.13. *Is the following equation exact?*

$$3x^2y^2dx + 6x^3ydy = 0 \tag{4.40}$$

Solution The given equation is $3x^2y^2dx + 6x^3ydy = 0$.

Comparing with $M(x,y)dx + N(x,y)dy = 0$, we get

$$M(x,y) = 3x^2y^2, \quad N(x,y) = 6x^3y,$$

$$M_y(x,y) = 6x^2y, \quad N_x(x,y) = 18x^2y$$

$$\therefore M_y(x,y) \neq N_x(x,y), \ \forall \ (x,y).$$

Therefore, the given equation is not exact.

Solution: The given equation is $6x^2y\,dx + 4x^3y\,dy = 0$.

Comparing with $M(x,y)dx + N(x,y)dy = 0$, we get

$$M(x,y) = 6x^2y, \quad N(x,y) = 4x^3y,$$

and

$$M_y(x,y) = N_x(x,y) = 12x^2y, \quad \forall(x,y).$$

The given equation is exact. We must find a function F such that

$$F_x(x,y) = M(x,y) = 6x^2y, \quad F_y(x,y) = 4x^3y. \tag{4.41}$$

To find F, integrating $F_x(x,y) = 6x^2y$, w.r.t x, we get

$$F(x,y) = 2x^3y + \phi(y) \tag{4.42}$$

where $\phi(y)$ is constant of integration. To determine ϕ so that F also satisfies (4.41), assume that ϕ is differentiable and differentiate F with respect to y.

Differentiating w.r.t y, we get

$$F_y(x,y) = 4x^3y + \phi'(y).$$

Comparing this with (4.41) shows that

$$\phi'(y) = 0.$$

Therefore, $\phi(y)$ is constant and it is taken to be zero.

Substituting this into (4.42) yields $F(x,y) = 2x^3y^2$.

Therefore, $2x^3y^2 = c$ is an implicit solution of given equation.

Example 4.15. *Solve*

$$(3y\cos x + 4xe^x + 2x^2e^x)dx + (3\sin x + 3)dy = 0. \tag{4.43}$$

Solution: The given equation is $(3y\cos x + 4xe^x + 2x^2e^x)dx + (3\sin x + 3)dy = 0$.

Comparing with $M(x,y)dx + N(x,y)dy = 0$, we get

$$M(x,y) = 3y\cos x + 4xe^x + 2x^2e^x, \quad N(x,y) = 3\sin x + 3,$$

and

$$M_y(x,y) = N_x(x,y) = 3\cos x, \quad \forall(x,y).$$

The given equation is exact. We must find a function F such that

$$F_x(x,y) = M(x,y) = 3y\cos x + 4xe^x + 2x^2e^x, \quad F_y(x,y) = 3\sin x + 3. \tag{4.44}$$

To find F, integrating $F_y(x,y) = 3\sin x + 3$, w.r.t y, we get where $\psi(x)$ is constant of integration. To determine $\psi(x)$ so that F also satisfies (4.44), assume that $\psi(x)$ is differentiable and differentiate F with respect to x.

Differentiating w.r.t. x, we get

$$F_x(x, y) = 3y \cos x + \psi'(x).$$

Comparing this with (4.44) gives

$$\psi'(x) = 4xe^x + 2x^2 e^x.$$

Integrating, we get

$$\psi(x) = 2x^2 e^x.$$

Putting in (4.45), we get

$$F(x, y) = 3y \sin x + 3y + 2x^2 e^x.$$

Therefore, $3y \sin x + 3y + 2x^2 e^x = c$ is an explicit solution of given equation.

Example 4.16. *Solve*

$$\left(\frac{1}{x} + 2x\right) dx + \left(\frac{1}{y} + 2y\right) dy = 0. \tag{4.46}$$

Solution: The given equation is $(\frac{1}{x} + 2x)dx + (\frac{1}{y} + 2y)dy = 0$.

Comparing with $M(x, y)dx + N(x, y)dy = 0$, we get

$$M(x, y) = \frac{1}{x} + 2x, \quad N(x, y) = \frac{1}{y} + 2y,$$

and

$$M_y(x, y) = N_x(x, y) = 0, \ \forall (x, y).$$

The given equation is exact. We must find a function F such that

$$F_x(x, y) = M(x, y) = \frac{1}{x} + 2x, \ F_y(x, y) = \frac{1}{y} + 2y. \tag{4.47}$$

To find F, integrating $F_x(x, y) = \frac{1}{x} + 2x$, w.r.t x, we get

$$F(x, y) = \log x + x^2 + \phi(y) \tag{4.48}$$

where $\phi(y)$ is constant of integration. To determine ϕ so that F also satisfies (4.47), assume that ϕ is differentiable and differentiate F with respect to y.

Differentiating w.r.t y, we get

$$F_y(x, y) = \phi'(y).$$

Comparing this with (4.47) shows that

$$\phi'(y) = \frac{1}{y} + 2y.$$

Therefore, $\phi(y) = \log y + y^2$, substituting this into (4.48) yields

$$F(x, y) = \log x + x^2 + \log y + y^2.$$

Therefore, $x^2 + y^2 + \log xy = c$ is an implicit solution of given equation.

Example 4.17. *Find all function M such that the equation $M(x, y)dx + (x^2 - y^2)dy = 0$ is exact.*

Solution: The differential equation $M(x, y)dx + N(x, y)dy = 0$ is exact if and only if

$$M_y(x, y) = N_x(x, y).$$

Here, $N(x, y) = x^2 - y^2$. Therefore,

$$M_y(x, y) = N_x(x, y) = 2x$$

Integrating, we get

$$M(x, y) = 2xy.$$

Example 4.18. *Find all function N such that the equation $(\ln xy + 2y \sin x)dx + N(x, y)dy = 0$ is exact.*

Solution: The differential equation $M(x, y)dx + N(x, y)dy = 0$ is exact if and only if

$$M_y(x, y) = N_x(x, y).$$

Here, $M(x, y) = \ln xy + 2y \sin x$. Therefore,

$$N_x(x, y) = M_y(x, y) = \frac{1}{xy}.x + 2 \sin x$$

$$N_x(x, y) = \frac{1}{y} + 2 \sin x$$

Integrating, we get

$$N(x, y) = \frac{x}{y} - 2 \cos x.$$

Remark 4.2. *Suppose all second partial derivatives of $F = F(x, y)$ are continuous and*

$$F_{xx} + F_{yy} = 0$$

on an open rectangle $\mathbb{R}$. A function with these properties is said to be harmonic. *If*

$$-F_y dx + F_x dy = 0$$

is exact on $\mathbb{R}$, and therefore there is a function G such that $G_x = -F_y$ and $G_y = F_x$ in $\mathbb{R}$ then a function G is said to be a harmonic conjugate *of F.*

Example 4.19. *Verify the function $F(x, y) = x^2 - y^2$ is harmonic and if so then find its harmonic*

Solution: Here, $F(x, y) = x^2 - y^2$.

$$F_x = 2x, \ F_y = -2y$$

$$F_{xx} = 2, \ F_{yy} = -2$$

$$F_{xx} + F_{yy} = 0$$

Therefore, F is harmonic.

The harmonic conjugate of a function F is

$$G_x = -F_y = 2y$$

$$\therefore G(x, y) = 2xy.$$

Hence, the harmonic conjugate of a function F is $2xy$.

Exercise 4.2

1. Determine which of the following equations are exact, if it is exact then solve.

    i. $14x^2y^3dx + 21x^2y^2dy = 0$

    ii. $(2x - 2y^2)dx + (12y^2 - 4xy)dy = 0$

    iii. $(x + y)^2dx + (x + y)^2dy = 0$

    iv. $(4x + 7y)dx + (3x + 4y)dy = 0$

    v. $(-2y^2 sinx + 3y^3 - 2x)dx + (4ycosx + 9xy^2)dy = 0$

    vi. $(2x + y)dx + (2y + 2x)dy = 0$

    vii. $(3x^2 + 2xy + 4y^2)dx + (x^2 + 8xy + 18y)dy = 0$

    viii. $(2x^2 + 8xy + y^2)dx + (2x^2 + xy^3/3)dy = 0$

    ix. $(\frac{1}{x} + 2x)dx + (\frac{1}{y} + 2y)dy = 0$

    x. $(ysinxy + xy^2cosxy)dx + (xsinxy + xy^2cosxy)dy = 0$

    xi. $(\frac{x}{(x^2+y^2)^{3/2}})dx + (\frac{y}{(x^2+y^2)^{3/2}})dy = 0$

    xii. $(e^x(x^2y^2 + 2xy^2) + 6x)dx + (2x^2ye^x + 2)dy = 0$

    xiii. $(x^2e^{x^2+y}(2x^2 + 3) + 4x)dx + (x^3e^{x^2+y} - 12y^2)dy = 0$

    xiv. $(e^{xy}(x^4y + 4x^3) + 3y)dx + (x^5e^{xy} + 3x)dy = 0$

    xv. $(3x^2cosxy - x^3ysinxy + 4x)dx + (8y - x^4sinxy)dy = 0$

2. Solve the following initial value problem.

   ii. $(-4y\cos x + 4\sin x\cos x + \sec^2 x)dx + (4y - 4\sin x)dy = 0,\ \ y(\pi/4) = 0$

   iii. $(y^3 - 1)e^x dx + 3y^2(e^x + 1)dy = 0,\ \ y(0) = 0$

   iv. $(\sin x - y\sin x - 2\cos x)dx + (\cos x)dy = 0,\ \ y(0) = 1$

   v. $(2x - 1)(y - 1)dx + (x + 2)(x - 3)dy = 0,\ \ y(1) = -1$

3. Find all function $M(x, y)$ such that the equation is exact.

   i. $M(x, y)dx + 2xy\sin x\cos y\, dy = 0$

   ii. $M(x, y)dx + (e^x - e^y\sin x)dy = 0$

4. Find all function $N(x, y)$ such that the equation is exact.

   i. $(x^3y^2 + 2xy + 3y^2)dx + N(x, y)dy = 0$

   ii. $(x\sin x + y\sin y)dx + N(x, y)dy = 0$

5. Verify that the following functions are harmonic and find all their harmonic conjugates.

   i. $e^x\cos y,$ ii) $x^3 - 3xy^2,$ iii) $\cos x\cosh y,$ iv) $\sin x\cosh y.$

4.3   Integrating Factors

In the previous section we saw that if M, N, M_y and N_x are continuous and $M_y = N_x$ on an open rectangle $\mathbb{R}$ then

$$M(x, y)dx + N(x, y)dy = 0 \tag{4.49}$$

is exact on $\mathbb{R}$. Sometimes an equation that is not exact can be made exact by multiplying it by an appropriate factor.

For example,

$$(3x + 2y^2)dx + 2xy\, dy = 0$$

is not exact, since $M_y(x, y) = 4y \neq N_x(x, y) = 2y.$

However, multiplying the given equation by x yields

$$(3x^2 + 2xy^2)dx + 2x^2y\, dy = 0$$

which is now exact, since $M_y(x, y) = N_x(x, y) = 4xy$, with the new $M(x, y)$ and $N(x, y)$ functions. Solving this renewed exact equation by the procedure given in Section 4.2 yields the implicit solution

$$x^3 + x^2y^2 = c.$$

A function $\mu = \mu(x, y)$ is an *integrating factor* for (4.49) if is exact. If we know an integrating factor μ for (4.49), we can solve the exact equation (4.50) by the method of Section 4.2.

Finding Integrating Factors

By applying theorem (4.2) with M and N replaced by μM and μN, we see that (4.50) is exact on an open rectangle $\mathbb{R}$ if μM, μN, $(\mu M)_y$ and $(\mu N)_x$ are continuous and

$$\frac{\partial}{\partial y}(\mu M) = \frac{\partial}{\partial N} \quad \text{or equivalently } \mu_y M + \mu M_y = \mu_x N + \mu N_x$$

on $\mathbb{R}$. It is better to rewrite the last equation as

$$\mu(M_y - N_x) = \mu_x N - \mu_y M, \tag{4.51}$$

which reduces to the known result for exact equations; that is, if $M_y = N_x$ then (4.51) holds with $\mu = 1$, so equation (4.49) is exact.

We shall now show that (4.51) is useful if we restrict our search to integrating factors that are products of a function of x and a function of y; that is, $\mu(x, y) = P(x)Q(y)$.

We are not saying that *every* equation $M\,dx + N\,dy = 0$ has an integrating factor of this form; rather, we are saying that *some* equations have such integrating factors.

We will now develop a way to determine whether a given equation has such an integrating factor, and a method for finding the integrating factor in this case.

If $\mu(x, y) = P(x)Q(y)$, then $\mu_x(x, y) = P'(x)Q(y)$ and $\mu_y(x, y) = P(x)Q'(y)$, so (4.51) becomes

$$P(x)Q(y)(M_y - N_x) = P'(x)Q(y)N - P(x)Q'(y)M, \tag{4.52}$$

or, after dividing throughout by $P(x)Q(y)$, we get

$$(M_y - N_x) = \frac{P'(x)}{P(x)}N - \frac{Q'(y)}{Q(y)}M. \tag{4.53}$$

Now let

$$p(x) = \frac{P'(x)}{P(x)} \quad \text{and } q(y) = \frac{Q'(y)}{Q(y)},$$

so (4.53) becomes

$$(M_y - N_x) = p(x)N - q(y)M. \tag{4.54}$$

We obtained (4.54) by *assuming* that $M\,dx + N\,dy = 0$ has an integrating factor $\mu(x, y) = P(x)Q(y)$. However, we can now view (4.53) differently:

If there are functions $p = p(x)$ and $q = q(y)$ that satisfy (4.54) and we define

$$P(x) = \pm e^{\int p(x)dx} \quad \text{and } Q(y) = \pm e^{\int q(y)dy}, \tag{4.55}$$

then reversing the steps that led from equations (4.52) to (4.54) shows that $\mu(x, y) = P(x)Q(y)$ is

(4.55) to be zero and choose the signs conveniently so the integrating factor has the simplest form. There's no simple general method for ascertaining whether functions $p = p(x)$ and $q = q(y)$ satisfying (4.54) exist. However, the next theorem gives simple sufficient conditions for the given equation to have an integrating factor that depends on only one of the independent variables x and y, and for finding an integrating factor in this case.

Theorem 4.3. *Let M, N, M_y and N_x be continuous on an open rectangle $\mathbb{R}$. Then*

(a). If $\dfrac{(M_y - N_x)}{N}$ is independent of y on $\mathbb{R}$ and we define

$$p(x) = \frac{M_y - N_x}{N},$$

then

$$\mu(x) = \pm e^{\int p(x)dx} \tag{4.56}$$

is an integrating factor for

$$M(x,y)dx + N(x,y)dy = 0 \tag{4.57}$$

on $\mathbb{R}$.

(b). If $\dfrac{(N_x - M_y)}{M}$ is independent of x on $\mathbb{R}$ and we define

$$q(y) = \frac{N_x - M_y}{M},$$

then

$$\mu(y) = \pm e^{\int q(y)dy} \tag{4.58}$$

is an integrating factor for equation (4.57) on $\mathbb{R}$.

Proof (a) If $\dfrac{(M_y - N_x)}{N}$ is independent of y, then

$$(M_y - N_x) = p(x)N - q(y)M$$

holds with

$$p = \frac{(M_y - N_x)}{N}, \quad \text{and } q \equiv 0.$$

Therefore,

$$P(x) = \pm e^{\int p(x)dx} \ \text{ and } \ Q(y) = \pm e^{\int q(y)dy} = \pm e^0 = \pm 1.$$

So (4.56) is an integrating factor for (4.57) on $\mathbb{R}$.

(b) If $\dfrac{(N_x - M_y)}{M}$ is independent of x then equation (4.54) holds with $q = \dfrac{(N_x - M_y)}{M}$ and $p \equiv 0$, and a similar argument shows that (4.58) is an integrating factor for (4.57) on $\mathbb{R}$.

Example 4.20. *Find an integrating factor for the equation*

$$(2xy^3 - 2x^3y^3 - 4xy^2 + 2x)dx + (3x^2y^2 + 4y)dy = 0 \tag{4.59}$$

and solve the equation.

Solution: The given equation is $(2xy^3 - 2x^3y^3 - 4xy^2 + 2x)dx + (3x^2y^2 + 4y)dy = 0$.
Comparing with $M(x, y)dx + N(x, y)dy = 0$, we get

$$M(x, y) = 2xy^3 - 2x^3y^3 - 4xy^2 + 2x \ \ N(x, y) = 3x^2y^2 + 4y,$$

and

$$M_y(x, y) = 6xy^2 - 6x^3y^2 - 8xy, \quad N_x(x, y) = 6xy^2.$$

$$M_y(x, y) \neq N_x(x, y).$$

The given equation is not exact. Therefore, we obtain the integrating factor. Now,

$$\frac{M_y - N_x}{N} = -\frac{6x^3y^2 + 8xy}{3x^2y^2 + 4y} = -2x$$

is independent of y. Therefore, above theorem 4.3(a) applies with $p(x) = -2x$.
Since

$$\int p(x)dx = -\int 2xdx = -x^2.$$

Hence, $\mu(x) = e^{-x^2}$ is an integrating factor.

Multiplying (4.59) by μ yields the exact equation

$$e^{-x^2}(2xy^3 - 2x^3y^3 - 4xy^2 + 2x)dx + e^{-x^2}(3x^2y^2 + 4y)dy = 0 \tag{4.60}$$

To solve this equation, we must find a function F such that

$$F_x(x, y) = e^{-x^2}(2xy^3 - 2x^3y^3 - 4xy^2 + 2x) \tag{4.61}$$

and

$$F_y(x, y) = e^{-x^2}(3x^2y^2 + 4y) \tag{4.62}$$

Integrating (4.62) with respect to y yields

$$F(x, y) = e^{-x^2}(x^2y^3 + 2y^2) + \psi(x). \tag{4.63}$$

Differentiating this with respect to x yields

$$F_x(x, y) = e^{-x^2}(2xy^3 - 2x^3y^3 - 4xy^2) + \psi'(x).$$

Comparing this with (4.61) shows that $\psi'(x) = 2xe^{-x^2}$.
Therefore, we can let $\psi(x) = -e^{-x^2}$ in (4.63) and conclude that

$$e^{-x^2}(y^2(x^2y + 2) - 1) = c$$

Example 4.21. *Find an integrating factor for the*

$$(2xy^3)dx + (3x^2y^2 + x^2y^3 + 1)dy = 0 \tag{4.64}$$

and solve the equation.

Solution In the given equation we have

$$M = 2xy^3, \quad N = 3x^2y^2 + x^2y^3 + 1,$$

and

$$M_y - N_x = 6xy^2 - (6xy^2 + 2xy^3) = -2xy^3.$$

Therefore, (4.64) is not exact.

Moreover,

$$\frac{M_y - N_x}{N} = -\frac{2xy^3}{3x^2y^2 + x^2y^2 + 1}$$

is not independent of y, so theorem $4.3(a)$ does not apply.

However, theorem $4.3(b)$ does apply, since

$$\frac{N_x - M_y}{M} = \frac{2xy^3}{2xy^3} = 1$$

is independent of x, so we can take $q(y) = 1$.

Since

$$\int q(y)dy = \int dy = y,$$

$\mu(y) = e^y$ is an integrating factor.

Multiplying (4.64) by μ yields the exact equation

$$2xy^3e^ydx + (3x^2y^2 + x^2y^3 + 1)e^ydy = 0. \tag{4.65}$$

To solve this equation, we must find a function F such that

$$F_x(x, y) = 2xy^3e^y. \tag{4.66}$$

and

$$F_y(x, y) = (3x^2y^2 + x^2y^3 + 1)e^y. \tag{4.67}$$

Integrating (4.66) with respect to x yields

$$F(x, y) = x^2y^3e^y + \phi(y). \tag{4.68}$$

Differentiating this with respect to y yields

$$F_y = (3x^2y^2 + x^2y^3)e^y + \phi'(y),$$

and comparing this with (4.67) shows that $\phi'(y) = e^y$.

Therefore, we set $\phi(y) = e^y$ in (4.68) and conclude that

$$(x^2 y^3 + 1)e^y = c$$

is an implicit solution of (4.65).

It is also an implicit solution of (4.64).

Note: Theorem 4.3 does not apply in the next example, but the more general argument that led to theorem 4.3 provides an integrating factor.

Example 4.22. *Find an integrating factor for*

$$(3xy + 6y^2)dx + (2x^2 + 9xy)dy = 0 \tag{4.69}$$

and solve the equation.

Solution In the given equation we have

$$M = 3xy + 6y^2, \quad N = 2x^2 + 9xy,$$

and

$$M_y - N_x = (3x + 12y) - (4x + 9y) = -x + 3y.$$

Therefore,

$$\frac{M_y - N_x}{M} = \frac{-x + 3y}{3xy + 6y^2} \quad and \quad \frac{N_x - M_y}{N} = \frac{x - 3y}{2x^2 + 9xy}.$$

So, theorem 4.3 does not apply.

Following the more general argument that led to theorem 4.3, we look for functions $p = p(x)$ and $q = q(y)$ such that

$$M_y - N_x = p(x)N - q(y)M;$$

that is,

$$-x + 3y = p(x)(2x^2 + 9xy) - q(y)(3xy + 6y^2).$$

Since the left side contains only first degree terms in x and y, we rewrite this equation as

$$xp(x)(2x + 9y) - yq(y)(3x + 6y) = -x + 3y.$$

This will be an identity if

$$xp(x) = A \quad and \quad yq(y) = B, \tag{4.70}$$

where A and B are constants such that

$$-x + 3y = A(2x + 9y) - B(3x + 6y),$$

or, equivalently,

Equating the coefficients of x and y on both sides shows that the last equation holds for all (x, y) if

$$2A - 3B = -1$$

$$9A - 6B = 3$$

, which has the solution $A = 1$, $B = 1$.

Therefore, (4.70) implies that

$$p(x) = \frac{1}{x} \ \ and \ \ q(y) = \frac{1}{y}.$$

Since

$$\int p(x)dx = \ln|x| \ \ and \ \ \int q(y)dy = \ln|y|.$$

We can let $P(x) = x$ and $Q(y) = y$; hence, $\mu(x, y) = xy$ is an integrating factor.

Multiplying (4.69) by μ yields the exact equation

$$(3x^2y^2 + 6xy^3)dx + (2x^3y + 9x^2y^2)dy = 0.$$

The reader may use the method of Section 4.2 to show that this method has an implicit solution given by

$$x^3y^2 + 3x^2y^3 = c. \tag{4.71}$$

which is also an implicit solution of (4.69).

Illustrative Examples

Example 4.23. *Find an integrating factor for the*

$$(5xy + 2y + 5)dx + (2x)dy = 0$$

and solve the equation.

Solution In the given equation we have

$$M = 5xy + 2y + 5, \ \ N = 2x,$$

where $M_y = 5x + 2$ and $N_x = 2$. Since $M_y \neq N_x$, that is,

$$M_y - N_x = 5x \neq 0,$$

so given equation is not exact. However,

$$\frac{M_y - N_x}{N} = \frac{5x}{2x} = \frac{5}{2}$$

is independent of y, so Theorem 4.3(a) is applicable and so we can take $p(x) = \frac{5}{2}$. Since

$$\int \qquad \int \frac{5}{} \qquad \frac{5x}{}$$

$\mu(x) = e^{\frac{5x}{2}}$ is an integrating factor. Multiplying the given equation by μ yields the exact equation

$$e^{\frac{5x}{2}}(5xy + 2y + 5)dx + (2x)e^{\frac{5x}{2}}dy = 0$$

To solve this equation, we must find a function F such that

$$F_x(x,y) = e^{\frac{5x}{2}}(5xy + 2y + 5).$$

and

$$F_y(x,y) = (2x)e^{\frac{5x}{2}}.$$

Integrating $F_x(x,y)$ with respect to x yields

$$\int F_x dx = F(x,y) = \int e^{\frac{5x}{2}}(5xy + 2y + 5)\ dx + \phi(y)$$

$$= \int (5yxe^{\frac{5x}{2}} + 2ye^{\frac{5x}{2}} + 5e^{\frac{5x}{2}})\ dx + \phi(y)$$

$$= 5y\left[xe^{\frac{5x}{2}}\frac{2}{5} - \frac{2}{5}e^{\frac{5x}{2}}\frac{2}{5}\right] + 2ye^{\frac{5x}{2}}\frac{2}{5} + 5e^{\frac{5x}{2}}\frac{2}{5} + \phi(y)$$

$$= 2xye^{\frac{5x}{2}} - \frac{4y}{5}e^{\frac{5x}{2}} + \frac{4y}{5}e^{\frac{5x}{2}} + 2e^{\frac{5x}{2}} + \phi(y)$$

$$ie.,\ \ F(x,y) = 2xye^{\frac{5x}{2}} + 2e^{\frac{5x}{2}} + \phi(y)$$

Differentiating this with respect to y yields $F_y = 2xe^{\frac{5x}{2}} + \phi'(y)$, and comparing this with above $F_y(x,y)$ shows that $2xe^{\frac{5x}{2}} + \phi'(y) = 2xe^{\frac{5x}{2}}$ implies $\phi'(y) = 0$. Therefore $\phi(y) = c$, for c constant and substituting in $F(x,y)$ we conclude that

$$2xye^{\frac{5x}{2}} + 2e^{\frac{5x}{2}} = c$$

is solution of given non-exact differential equation.

Example 4.24. *Find an integrating factor for the*

$$(6xy^2 + 2y)dx + (12x^2y + 6x + 3)dy = 0$$

and solve the equation.

Solution In the given equation we have

$$M = 6xy^2 + 2y,\ \ N = 12x^2y + 6x + 3,$$

where $M_y = 12xy + 2$ and $N_x = 24xy + 6$. Since $M_y \neq N_x$, that is,

$$M_y - N_x = -12xy - 4 \neq 0,$$

so given equation is not exact. Moreover,

$$M_y - N_x \qquad -12xy - 4 \qquad -12xy - 4$$

is not independent of y, so Theorem 4.3(a) is not applicable. However,

$$\frac{N_x - M_y}{M} = \frac{12xy + 4}{6xy^2 + 2y} = \frac{2(6xy + 2)}{y(6xy + 2)} = \frac{2}{y}$$

is independent of x, so Theorem 4.3(b) is applicable. So we can take $q(y) = \frac{2}{y}$. Since

$$\int q(y)\, dy = \int \frac{2}{y}\, dy = 2 \log y = \log y^2,$$

$\mu(x) = e^{\log y^2} = y^2$ is an integrating factor. Multiplying the given equation by μ yields the exact equation

$$y^2(6xy^2 + 2y)dx + y^2(12x^2y + 6x + 3)dy = 0$$

that is,

$$(6xy^4 + 2y^3)dx + (12x^2y^3 + 6xy^2 + 3y^2)dy = 0$$

To solve this equation, we must find a function F such that

$$F_x(x, y) = 6xy^4 + 2y^3$$

and

$$F_y(x, y) = 12x^2y^3 + 6xy^2 + 3y^2.$$

Integrating $F_x(x, y)$ with respect to x yields

$$\int F_x dx = F(x, y) = \int 6xy^4 + 2y^3 \ dx + \phi(y)$$

$$= 6\frac{x^2}{2}y^4 + 2xy^3 + \phi(y)$$

$$ie., \quad F(x, y) = 3x^2y^4 + 2xy^3 + \phi(y)$$

Differentiating this with respect to y yields $F_y = 12x^2y^3 + 6xy^2 + \phi'(y)$, and comparing this with above $F_y(x, y)$ shows that $12x^2y^3 + 6xy^2 + \phi'(y) = 12x^2y^3 + 6xy^2 + 3y^2$ implies $\phi'(y) = 3y^2$. Therefore $\phi(y) = y^3 + c$ and substituting this in $F(x, y)$ concludes that

$$3x^2y^4 + 2xy^3 + y^3 = c$$

is solution of given non-exact differential equation.

Example 4.25. *Find an integrating factor for the following differential equation and solve.*

$$(cos \ x \ cos \ y)dx + (sin \ x \ cos \ y - sin \ x \ sin \ y + y)dy = 0$$

Solution In the given equation we have

$$M = (cos \ x \ cos \ y), \quad N = (sin \ x \ cos \ y - sin \ x \ sin \ y + y),$$

where $M_y = -\cos x \sin y$ and $N_x = \cos x \cos y - \cos x \sin y$. Since $M_y \neq N_x$, the given equation is not exact. However,

$$\frac{N_x - M_y}{M} = \frac{\cos x \cos y - \cos x \sin y + \cos x \sin y}{\cos x \cos y} = 1,$$

is independent of x, so Theorem 4.3(b) is applicable. So we can take $q(y) = 1$. Since

$$\int q(y)\,dy = \int 1\,dy = y,$$

$\mu(x) = e^y$ is an integrating factor. Multiplying the given equation by μ yields the exact equation

$$e^y(\cos x \cos y)dx + e^y(\sin x \cos y - \sin x \sin y + y)dy = 0$$

To solve this equation, we must find a function F such that

$$F_x(x, y) = e^y(\cos x \cos y)$$

and

$$F_y(x, y) = e^y(\sin x \cos y - \sin x \sin y + y).$$

Integrating $F_x(x, y)$ with respect to x yields

$$\int F_x dx = F(x, y) = \int e^y(\cos x \cos y)\,dx + \phi(y)$$
$$= e^y \cos y \sin x + \phi(y)$$
$$ie., \quad F(x, y) = e^y \cos y \sin x + \phi(y)$$

Differentiating this with respect to y yields $F_y = -e^y \sin x \sin y + e^y \cos y \sin x + \phi'(y)$, and comparing this with above $F_y(x, y)$ shows that $-e^y \sin x \sin y + e^y \cos y \sin x + \phi'(y) = e^y \sin x \cos y - e^y \sin x \sin y + e^y y$ implies $\phi'(y) = ye^y$. Therefore $\phi(y) = e^y(y - 1) + c$ and substituting this in $F(x, y)$ concludes that

$$e^y \cos y \sin x + e^y(y - 1) = c$$

is solution of the given non-exact differential equation.

Example 4.26. *Find an integrating factor and solve the following differential equation.*

$$(x^4 y^3 + y)dx + (x^5 y^2 - x)dy = 0$$

Solution In the given equation we have

$$M = x^4 y^3 + y, \quad N = x^5 y^2 - x,$$

where $M_y = 3x^4 y^2 + 1$ and $N_x = 5x^4 y^2 - 1$. Since $M_y \neq N_x$, the given equation is not exact. However,

$$M_y - N_x \qquad -2x^4 y^2 + 2 \qquad -2(x^4 y^2 - 1) \qquad 2$$

is independent of y, so Theorem $4.3(a)$ is applicable and so we can take $p(x) = -\frac{2}{x}$. Since

$$\int p(x)dx = \int -\frac{2}{x}dx = -2\log x = \log x^{-2},$$

$\mu(x) = e^{\log x^{-2}} = x^{-2}$ is an integrating factor. Multiplying the given equation by μ yields the exact equation

$$x^{-2}(x^4 y^3 + y)dx + x^{-2}(x^5 y^2 - x)dy = 0$$

that is,

$$(x^2 y^3 + \frac{y}{x^2})dx + (x^3 y^2 - \frac{1}{x})dy = 0$$

To solve this equation, we must find a function F such that

$$F_x(x,y) = x^2 y^3 + \frac{y}{x^2}.$$

and

$$F_y(x,y) = x^3 y^2 - \frac{1}{x}.$$

Integrating $F_x(x,y)$ with respect to x yields

$$\int F_x(x,y)dx = F(x,y) = \int \left(x^2 y^3 + \frac{y}{x^2} \right) dx + \phi(y)$$

$$= \frac{x^3}{3} y^3 + y\left(\frac{-1}{x} \right) + \phi(y)$$

$$ie., \quad F(x,y) = \frac{x^3 y^3}{3} - \frac{y}{x} + \phi(y)$$

Differentiating this with respect to y yields $F_y = x^3 y^2 - \frac{1}{x} + \phi'(y)$, and comparing this with above $F_y(x,y)$ shows that $x^3 y^2 - \frac{1}{x} + \phi'(y) = x^3 y^2 - \frac{1}{x}$ implies $\phi'(y) = 0$. Therefore $\phi(y) = c$, for c constant and substituting in $F(x,y)$ we conclude that

$$\frac{x^3 y^3}{3} - \frac{y}{x} = c$$

is solution of the given non-exact differential equation.

Exercise 4.3

1. State the given statement is true or false and give justification to your answer.

   i. The differential equation $(2ye^{2x} - \sin y)dx + (e^{2x} - x\cos y)dy = 0$ is exact.

   ii. The differential equation $(y^x + \cos x)dx + 2xy^2 dy = 0$ is exact.

2. Determine whether the following differential equations is exact.

   i. $(y + 3x^2)dx + xdy = 0,$ ii. $(y^2 + \cos x)dx + (2xy + \sin xy)dy = 0,$

3. Find an integrating factor; that is a function of only one variable, and solve the given equation.

   i. $ydx - xdy = 0$ Ans: $y = c\,x$

   ii. $3x^2ydx + 2x^3dy = 0$ Ans: $y = c\,e^{-\frac{3\log(x)}{2}}$

   iii. $(5xy + 2y + 5)dx + 2xdy = 0$ Ans: $y = \left(c - e^{\frac{5\,x}{2}}\right)e^{-\frac{2\log(x)+5\,x}{2}}$

   iv. $(xy + x + 2y + 1)dx + (x + 1)dy = 0$ Ans: $y = \frac{e^{-x}\left(-(x-1)\,e^x - e^x + \%c\right)}{x+1}$

   v. $(27xy^2 + 8y^3)dx + (18x^2y + 12xy^2)dy = 0$ Ans: $4\,x^2\,y^3 + 9\,x^3\,y^2 = c$

   vi. $(6xy^2 + 2y)dx + (12x^2y + 6x + 3)dy = 0$ Ans: $3\,x^2\,y^4 + (2\,x + 1)\,y^3 = c$

   vii. $y^2dx + (xy^2 + 3xy + \frac{1}{y})dy = 0$ Ans: $(x\,y^3 + 1)\,e^y = c$

   viii. $(12x^3y + 24x^2y^2)dx + (9x^4 + 32x^3y + 4y)dy = 0$ Ans: $(8\,x^3 + 1)\,y^4 + 3\,x^4\,y^3 = c$

   ix. $(x^2y + 4xy + 2y)dx + (x^2 + x)dy = 0$ Ans: $y = \frac{\%c\,e^{-x}}{x^2\,(x+1)}$

   x. $(-y)dx + (x^4 - x)dy = 0$ Ans: $y = \frac{\%c\,e^{\frac{\log\left(x^2+x+1\right)}{3} + \frac{\log(x-1)}{3}}}{x}$

   xi. $(\cos x\cos y)dx + (\sin x\cos y - \sin x\sin y + y)dy = 0$ Ans: $\sin(x)\,e^y\cos(y) + (y - 1)\,e^y = c$

   xii. $(2xy + y^2)dx + (2xy + x^2 - 2x^2y^2 - 2xy^3)dy = 0$ Ans: $(x\,y^2 + x^2\,y)\,e^{-y^2} = c$

   xiii. $(y\sin y)dx + x(\sin y - y\cos y)dy = 0$ Ans: $\log(\sin(y)) - \log(y) = \log(x) + c.$

Maxima Lab Work

1. Use the commands $'diff(y, x)$, $'diff(y, x, 2)$, represent following ordinary differential equations.

   (a) $y' + 3y = 1,$

   (b) $y' + (\frac{1}{x} - 1)y = -\frac{2}{x},$

   (c) $y' + (\frac{2x}{1+x^2})y = \frac{e^{-x}}{1+x^2}$

   (d) $y' + \frac{1}{x}y = \frac{7}{x^2} + 3,$

   (e) $y' + \frac{4}{x-1}y = \frac{1}{(x-1)^5} + \frac{\sin x}{(x-1)^4},$

   (f) $y' + (\tan x)y = \cos x,$

   (g) $y' + (\sin 2x)y = e^{-\sin^2 x}.$

2. Find the integrating factor using maxima for following differential equations to be exact.

   (a) $ydx - xdy = 0$

   (b) $3x^2ydx + 2x^3dy = 0$

   (c) $2y^3dx + 3y^2dy = 0$

(e) $(xy + x + 2y + 1)dx + (x + 1)dy = 0$

(f) $(27xy^2 + 8y^3)dx + (18x^2y + 12xy^2)dy = 0$

(g) $(6xy^2 + 2y)dx + (12x^2y + 6x + 3)dy = 0$

(h) $y^2dx + (xy^2 + 3xy + \frac{1}{y})dy = 0$

(i) $(12x^3y + 24x^2y^2)dx + (9x^4 + 32x^3y + 4y)dy = 0$

(j) $(x^2y + 4xy + 2y)dx + (x^2 + x)dy = 0$

(k) $(-y)dx + (x^4 - x)dy = 0$

(l) $(\cos x \cos y)dx + (\sin x \cos y - \sin x \sin y + y)dy = 0$

(m) $(2xy + y^2)dx + (2xy + x^2 - 2x^2y^2 - 2xy^3)dy = 0$

(n) $(y\sin y)dx + x(\sin y - y \cos y)dy = 0$

3. Determine which of the following differential equations are exact.

(a) $6x^2y^2dx + 4x^3ydy = 0$

(b) $(3y\cos x + 4xe^x + 2x^2e^x)dx + (3\sin x + 3)dy = 0$

(c) $14x^2y^3dx + 21x^2y^2dy = 0$

(d) $(2x - 2y^2)dx + (12y^2 - 4xy)dy = 0$

(e) $(x + y)^2dx + (x + y)^2dy = 0$

(f) $(4x + 7y)dx + (3x + 4y)dy = 0$

(g) $(-2y^2\sin x + 3y^3 - 2x)dx + (4y\cos x + 9xy^2)dy = 0$

(h) $(2x + y)dx + (2y + 2x)dy = 0$

(i) $(3x^2 + 2xy + 4y^2)dx + (x^2 + 8xy + 18y)dy = 0$

(j) $(2x^2 + 8xy + y^2)dx + (2x^2 + xy^3/3)dy = 0$

(k) $(\frac{1}{x} + 2x)dx + (\frac{1}{y} + 2y)dy = 0$

(l) $(y\sin xy + xy^2\cos xy)dx + (x\sin xy + xy^2\cos xy)dy = 0$

(m) $(\frac{x}{(x^2+y^2)^{3/2}})dx + (\frac{y}{(x^2+y^2)^{3/2}})dy = 0$

(n) $(e^x(x^2y^2 + 2xy^2) + 6x)dx + (2x^2ye^x + 2)dy = 0$

(o) $(x^2e^{x^2+y}(2x^2 + 3) + 4x)dx + (x^3e^{x^2+y} - 12y^2)dy = 0$

(p) $(e^{xy}(x^4y + 4x^3) + 3y)dx + (x^5e^{xy} + 3x)dy = 0$

(q) $(3x^2\cos xy - x^3y\sin xy + 4x)dx + (8y - x^4\sin xy)dy = 0$

4. Using code ode2, solve the following initial value problem.

(b) $xy' + (1 + \frac{1}{\ln x})y = 0, \ y(e) = 1.$

(c) $xy' + (1 + x \cot x)y = 0, \ y(\pi/2) = 2.$

(d) $y' - (\frac{2x}{1+x^2})y = 0, \ y(0) = 2.$

(e) $y' + (\frac{k}{x})y = 0, \ y(1) = 3, (k \, constant).$

(f) $y' + (\tan k\,x)y = 0, \ y(0) = 2, (k \, constant).$

(g) $y' + 7y = e^{3x}, \ y(0) = 0$

(h) $y' = \frac{y^2 - 3xy - 5x^2}{x^2}, \ y(1) = -1$

(i) $x^2 y' = 2x^2 + y^2 + 4xy, \ y(1) = 1$

(j) $xyy' = 3x^2 + 4y^2, \ y(1) = \sqrt{3}$

(k) $(4x^3 y^2 - 6x^2 y - 2x - 3)dx + (2x^4 y - 2x^3)dy = 0, \ y(1) = 3$

(l) $(-4y\cos x + 4\sin x \cos x + \sec^2 x)dx + (4y - 4\sin x)dy = 0, \ y(\pi/4) = 0$

(m) $(y^3 - 1)e^x dx + 3y^2(e^x + 1)dy = 0, \ y(0) = 0$

(n) $(\sin x - y\sin x - 2\cos x)dx + (\cos x)dy = 0, \ y(0) = 1$

(o) $(2x - 1)(y - 1)dx + (x + 2)(x - 3)dy = 0, \ y(1) = -1.$

5. Plot a direction field and some integral curves on the indicated rectangular region.

(a) $x^2 y' = xy + x^2 + y^2; \ -8 \le x \le 8, -8 \le y \le 8$

(b) $xyy' = x^2 + 2y^2; \ -4 \le x \le 4, -4 \le y \le 4$

(c) $y' = \frac{2y^2 + x^2 e^{-(y/x)^2}}{2xy}; \ -8 \le x \le 8, -8 \le y \le 8$

(d) $y' - xy = x^3 y^3; \ -3 \le x \le 3, -2 \le y \le 2$

(e) $y' - y\frac{1+x}{3x} = y^4; \ -2 \le x \le 2, -2 \le y \le 2$